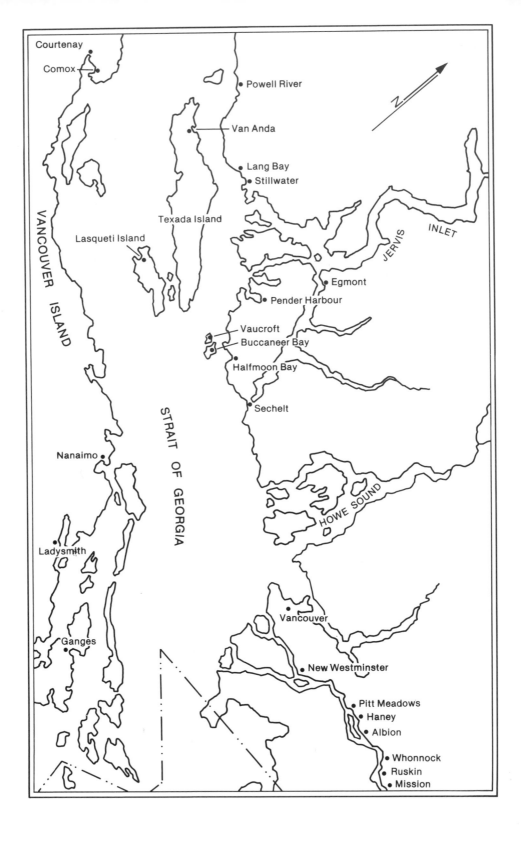

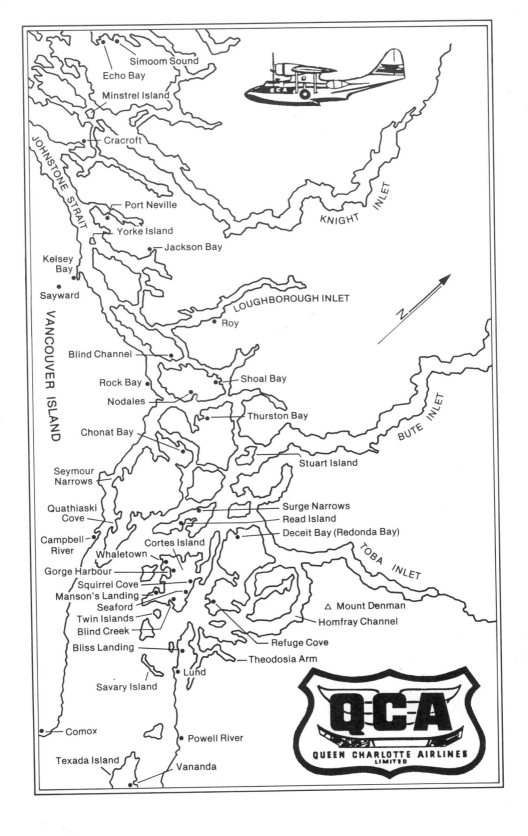

The Accidental Airline

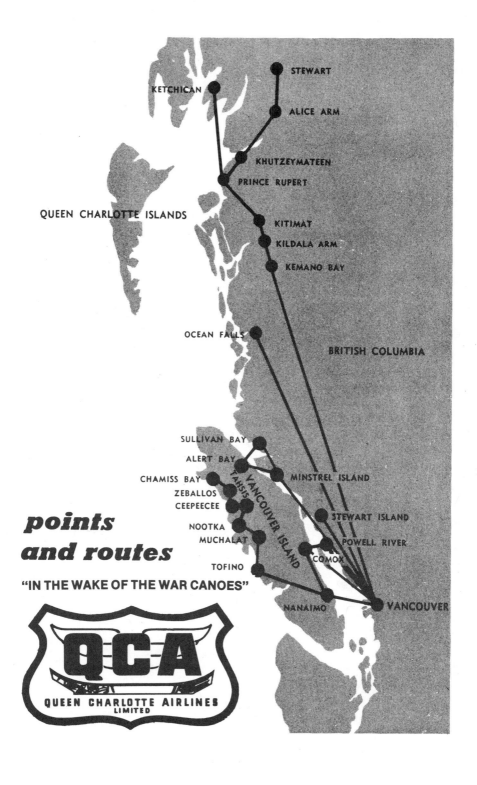

STEWART

KETCHICAN

ALICE ARM

KHUTZEYMATEEN

PRINCE RUPERT

QUEEN CHARLOTTE ISLANDS

KITIMAT

KILDALA ARM

KEMANO BAY

OCEAN FALLS

BRITISH COLUMBIA

SULLIVAN BAY

ALERT BAY

CHAMISS BAY

ZEBALLOS

CEEPEECEE

TAHSIS

VANCOUVER ISLAND

MINSTREL ISLAND

STEWART ISLAND

NOOTKA

MUCHALAT

POWELL RIVER

COMOX

TOFINO

points
and routes

"IN THE WAKE OF THE WAR CANOES"

NANAIMO

VANCOUVER

QCA

QUEEN CHARLOTTE AIRLINES
LIMITED

The Accidental Airline

SPILSBURY'S QCA

Howard White
Jim Spilsbury

Harbour Publishing

First paperback printing, 1994, second paperback printing 1996, third paperback printing, 1998, fourth paperback printing, 2004

Published by
Harbour Publishing Co. Ltd., P.O. Box 219, Madeira Park, BC V0N 2H0
www.harbourpublishing.com

Cover pastel by A.J. Spilsbury
Cover design by Fiona Macgregor
Printed and bound in Canada

Harbour Publishing acknowledges financial support from the Government of Canada through the Book Publishing Industry Development Program and the Canada Council for the Arts; and from the Province of British Columbia through the British Columbia Arts Council and the Book Publisher's Tax Credit through the Ministry of Provincial Revenue and the British Columbia Heritage Trust.

THE CANADA COUNCIL | LE CONSEIL DES ARTS
FOR THE ARTS | DU CANADA
SINCE 1957 | DEPUIS 1957

BRITISH
COLUMBIA
ARTS COUNCIL
Supported by the Province of British Columbia

Canadian Cataloguing in Publication Data

White, Howard, 1945–
 The accidental airline

ISBN 1-5017-097-X

 1. Queen Charlotte Airlines—History. 2. Airlines—British Columbia Vancouver—History. I. Spilsbury, Jim, 1905–2003 II. Title.
HE9815.Q43W54 1988 387.7'065'71133 C88-091380-0

Contents

ONE

The Non-Flyer

APART FROM AIR CANADA, which was created by an act of God, most airlines in Canada were started on shoestrings by overweening bushpilots. Grant McConachie started Yukon Southern, Max Ward started Wardair, and so on. Queen Charlotte Airlines, which later became Pacific Western Airlines, and still later Canadian International, was a different story. It was started on a shoestring rightly enough, but by a confirmed passenger, namely myself. I didn't learn to fly until quite well on in the game.

I guess I wanted to fly but I was too darn busy trying to keep things together to take the time. However, when the airline got well going and we had two or three hundred people there and twenty or thirty aircraft I got enough time off that I went to the Aero Club and took lessons in a Tiger Moth with Len Milne as instructor. I went up a few times, but I couldn't find any connection between the stick and what happened to the aeroplane. It seemed like there was a missing link there or something. I had a hell of a time of it and I was most unhappy.

The fifth time I was out there Len said, okay, next time you're on your own. Well, *no way* was I ready. I hadn't managed to land the thing yet without bouncing it or putting it on its nose. I never went back. It scared the heck out of me.

About a year later I was talking to Stan Sharpe at Brisbane Aviation and he took pity on me. He said, look, we won't charge you the full rate, come on over and carry on with your lessons. They had a different kind of aircraft. I learned then on an Aeronca

Tandem — the student sat forward and the instructor behind him. It was much easier to fly. Very forgiving. It made me wonder why they ever persisted in trying to teach people on Tiger Moths. They killed hundreds of people in Tiger Moths during the war. They were dreadful things it seemed to me, but the oldtimers swore by them.

My instructor this time around was a bit more of a psychologist. After we'd taken a few practice landings and takeoffs and we were waiting to go up again he said, you just keep on going like that, I'm going to go have a smoke. I was five hundred feet in the air before it dawned on me that he wasn't in the seat behind me. I was solo! I got my licence, I flew on wheels just long enough to get checked out on floats and from then on I used to fly a little four-place Stinson Station Wagon that the company had, CF-FYJ. I could putter up the coast on my own, go up to Savary Island with the family, and I really enjoyed it.

All in all I think I have less than 250 hours accumulated time as a pilot, but I could fill a book with all the engine failures I've had, prangs and near prangs I've got into. So when they came out with the new licences and everybody had to go back and rewrite, that's when I decided to hell with it, I had better things to do. It was just as well.

CF-FYJ survived all of my misadventures only to come to grief at the hands of a professional pilot, who left it moored under a wharf while the tide came in — a landlubber stunt I could never have committed in my worst moment of confusion. This afforded me such a delicious moment of relief from the sense of inferiority I normally suffered dealing with real pilots, it was almost worth the loss. We replaced FYJ with an another Stinson, CF-FFW, which I attached myself to with renewed confidence. This buoyant new period of my flying career lasted until the late afternoon of September 22, 1949. I still have the details in my logbook, so I can be precise about the date.

I was gung-ho to fly. All I needed was a legitimate excuse, and I figured I had it. Louis Potvin, our radio service manager, was up at Butedale repairing the radio-telephone for the Canadian Fishing Company, and needed transportation to Prince Rupert and Cumshewa Inlet in the Queen Charlottes for other radio jobs. I had planned a trip to Prince Rupert anyway, and told him I would pick him up on my way by. At the same time one of our Norseman aircraft was disabled at Maude and Oscar Johnson's logging camp in Belize Inlet, and our flight operations manager Johnny Hatch said I might as well do something useful and take a mechanic and spare cylinder pot up there, since it was all on my route. This, of course, gave me a great feeling of importance. I was already feeling more distinguished than usual since I had just bought myself a pair of dark glasses — airforce style — quite expensive, but definitely the in thing with pilots. Every pilot wore them, day and night. If you didn't have a pair you simply didn't look like a pilot. Now that I could fly an aeroplane I wanted to look the part. It was great. There was one slight hitch however; they tended to restrict vision somewhat, particularly if the light was not very good, but this disadvantage I was quite happy to accept in the interests of appearance.

It was a nice clear day and I took FFW off from Vancouver late in the afternoon with a QCA mechanic, "Scotty" Graham, his tool kit and the spare cylinder. I had my dark glasses on of course, and just to add to the impression of professionalism, I got out my new circular slide rule and worked out all the problems I had learned in class — air-speed, allowance for wind drift, corrected compass heading and all the rest of the garbage — even though visibility was so good I could just about see Belize Inlet from Vancouver. But it impressed the hell out of Scotty. It appeared that he had just recently started to take flying lessons, and his interest in all the techniques of flying was at peak. I took her up to about ten thousand feet and flew a direct course to the head of Seymour Inlet, passing well over all mountain tops en route. We had a good view of Mount Waddington and its associated glaciers, about twenty miles on our starboard hand. After about two hours we passed Mount Stephens directly under us, and could now see the entire length of Seymour Inlet, about fifty miles of it, looking like a very large and very long canal. I started our let-down procedure, using the circular slide rule to work out speed and rate of descent, calculating to put down about a mile short of the end of the inlet, where Oscar Johnson's logging camp was situated and where the Norseman awaited our arrival. I called Maude Johnson on the radio and gave her our ETA, about twenty minutes, and she reported no wind and the sea calm, but warned me to look out for their pet whale — don't hit him! Apparently this large whale had been hanging around the camp for some days. I thanked her and promised I would stay well clear of the whale. I throttled back and adjusted our glide to two hundred feet a minute, using the "Rate of Climb" instrument on the dash. Everything was perfect. Scotty was impressed all over again, and said so. Twenty minutes to put us on the water, just where we wanted to be.

It was now evening, and the sun was close to the horizon. As we descended into the trough of the inlet, below mountain-top level, the sun was blocked off by the mountains, and there being no clouds or scatter light, the cockpit went suddenly quite dark. Thanks to my new shades I could no longer read the instruments on the dash, but everything was going so well it didn't occur to me to take them off, and as we descended into denser air the rate of descent decreased and levelled off so we were flying almost straight and level, with still a hundred feet or so to descend. On top of this, we had a classic case of *glassy water* conditions which all float plane pilots are taught to recognize and respect. You cannot judge your distance to the water. It just looks like a huge mirror, which it is. I therefore adopted the recommended procedure and carried out a *glassy water approach* — throttle back, nose up — and feel your way down till your floats touch. I did all this according to the book, explaining it to Scotty, step by step.

The lights of the camp were now showing up ahead of us, and we were rapidly running out of space. Then the aircraft seemed to pitch ever so gently forward and level out. I figured the water was so smooth that we never felt the floats touch. Scotty swung around in his seat and reached for my hand to shake it. "Man," he said, "that's the smoothest landing I've ever had!" I gave him an appreciative smile in acknowledgement, and gently closed the throttle — at which the aircraft dropped about forty feet onto the water and bounced up in the air again. I felt the urge to open the throttle full and attempt to fly out of the stall, but common sense prevailed. We were much too close to the hillside, so I just sat tight and let her come down again, which she did in a very un-dignified manner, landing right on top of one of the log floats of the camp.

I took off my sunglasses to see where I was. The Stinson's propeller had come to a stop less than six inches away from a five-thousand-gallon gasoline storage tank. The pontoons were full length out of the water — almost undamaged. The gas-float on which we had landed was tied up alongside the cookhouse, where approximately forty men had been peacefully eating their supper. Our spectacular arrival brought every one of them out to stare in disbelief at us and our aeroplane, which had its starboard wing tip almost in the cookhouse door. On top of this, who should come stumbling out of the cookhouse with the rest of the crowd but "Uppy" Upson, the Assistant District Inspector of Civil Aviation for Canada, Western Division, who just happened to be spending a

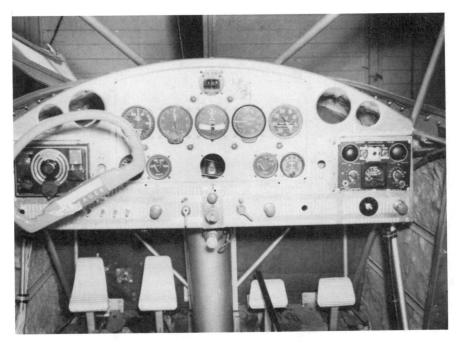

An aeroplane I could fly—sort of. Stinson controls.

On the second bounce I landed right at Maude and Oscar's cookhouse door.

I was the butt of many jokes that day. Here Maude Johnson purports to show the true meaning of "seat of the pants flying."

relaxing weekend away from all his aviation worries, hoping for a little quiet fishing!

Oscar Johnson rushed up to my window, where I was still contemplating the gasoline storage tank, and blurted, "Yumping Yesus, Yimmie, what happened? You scared the Beyesus out of us!" Then Maude rushed up, passed me a full glass of rye whiskey and asked if we were hurt. I tried to apologize for the disturbance to their supper, pointing out that in any case I had managed to miss the whale. I declined the whiskey with thanks, since I didn't drink in those days, so Maude took it back and downed it herself in two gulps.

Scotty examined the aircraft carefully and the only damage he could find was a very small puncture in the bottom of one of the pontoons where the toggle of a boom-chain had been protruding from the float log. He applied a small patch of aluminum with PK screws and poured in a little roofing pitch to make it watertight. All we had to do was get three or four men and skid it back into the water and I figured I would be on my way in the morning. But I hadn't figured on Uppy Upson. As senior government inspector on the scene, he decreed that the aircraft must be ferried back to Vancouver base without passengers, and by a properly licensed commercial pilot.

I never tried to fly with sunglasses again. In fact, I have never worn sunglasses again to this day.

I have been known to say that I became involved in the flying business purely by accident. Looking back, it often seemed that way, and in one sense it was true: I never expected what happened to happen. On the other hand, I'd had an interest in planes as early as I can remember. I could almost date when it started. It was while we were still living on the old family farm in Whonnock and we were in Vancouver to visit Mother's friends, the Burpees. Mr. Burpee was the head of Canadian Customs in Vancouver and when we got a parcel from England he always recognized it. He had four daughters — and from the time I was knee high we would stay at their place whenever we went to town. I was a little kid running around there and these four girls made a great fuss over me. I felt very important. But there was one thing I never forgave them for. One time when we were staying at their place, everybody decided to go out to Richmond to see a man take off in a flying machine.

I was dying to see this, but these girls decided that it would be better for me to stay in town and watch the Boy Scout parade, no doubt anticipating a crash. One of them was detailed off to oversee me while Mother and all the rest of them sallied out to Richmond. To add insult to injury, the Boy Scout parade never did materialize.

Possibly by way of placating me, somebody helped me make, out of matchsticks and bits of bamboo and butter paper, a model of one of these spindly biplanes that were considered excitingly progressive at the time. This was my pride and joy and I had it with me in the train going up to Whonnock when some older kid talked me into holding it out the window to see how it flew. When the wind hit, it disappeared in a puff, like when you blow on a dandelion.

I was about four or five, so this flight I was excited about would have to be the American Charles K. Hamilton, who made the first flight in western Canada at Minoru Park racetrack on Lulu Island March 25, 1910, amidst a lot of ballyhoo and with half the city watching.

So I was aware of flying from the very beginning. What kept me thinking about it was my cousin Rupert. If you have read my first book you will remember Uncle Ben Spilsbury, the second in the series of three Spilsbury brothers who originally left Derbyshire to settle in Whonnock in 1878. In 1889 Uncle Ben enticed my seventeen-year-old father, Ashton Wilmot Spilsbury, to leave his medical studies at Cambridge to take over the 360-acre Whonnock property, which Dad then cleared and farmed while Ben went into business with the famous industrialist R.V. Winch. Well, old Winch went broke in the Great Depression of 1893 and Uncle Ben had to take refuge back in Derbyshire, but Winch soon picked himself up and started up all over again, and he sent Uncle Ben a message to the effect that if he came back out, Winch would make him back all the money he'd lost. Ben came out with his two sons, Dick and Rupert, and my Aunt Edie. They had two daughters, but left them with all my other aunts in England to put through school and raise. I suppose they thought this rough country was no place for a young woman or something—the English never needed much of an excuse to start farming out their children. This would be just previous to the start of World War I. Winch did exactly as he promised. He got Ben to buy a lot of land in North Vancouver, which became very valuable, and in no time at all my uncle was financially independent. Winch himself went on to become a millionaire all over again, twice in a lifetime.

Uncle Ben never really did do anything. He'd hardly work around the house. He kept right on with the old-country style. They lived in North Vancouver, and I remember going to visit them with Dad. We would come down from Savary and check in at the Alcazar Hotel, then phone up and arrange to go over on the old North Van ferry and walk up Lonsdale to their place on Eighth Street. It was a strange household, to me. Uncle Ben was lord and master. He sat there with his mug of beer and was waited on hand and foot. My aunt Edie would hardly show her face. She would do all the housework and cooking, but she wouldn't even eat her meals with the men. She was a wonderful person, we all thought the world of her, and she blossomed out and became a person in her own right after Uncle Ben died. But while he was alive she followed the dictum that a woman's place is in the kitchen. The first time we went over there—this was after my dad lost the Whonnock farm and moved us to Savary Island—it got to be late in the afternoon and Aunt Edie said "Come into the dining room now and we'll have tea." There were biscuits and things, but I was trained not to eat between meals lest it spoil my supper, so I didn't take anything. But that was it. Supper never came. As true Brits, they didn't observe supper. Dad was caught on it too, so we took an early ferry back and on the way to the Alcazar stopped at the Trocadero Cafe and loaded up.

In my early days, after I got through school and was doing very little that was exciting, I used to get down to Vancouver about once a year, and when I did I would drop over to see my cousins, particularly Rupert. Rupert was always into something, frequently things that were somewhat questionable. He'd try anything. One of the things I remember was an outfit selling secondhand boats. I think it was called Pasco's. They had a very bad name. Rupert fitted in nicely. He took me down to show me some of these dilapidated old yachts in Coal Harbour, and this particular one had a number of pyrene fire extinguishers in brackets at strategic places around the cabin and cockpit. Rupert grabbed one and squirted a little bit out. "Recognize that?" he said. I sniffed it. It was pure gasoline. "They were all empty, and we had to fill them with something!" he said. He thought it was funny as hell.

The next thing, he started taking flying lessons out at the temporary airport at the racetrack in Richmond. There was still no permanent airport in Vancouver and they were doing business in a field beside the track. That was Vancouver's only airport until 1931, when they moved to Sea Island. Rupert was very involved with the flying scene at that time and travelling with a very fast gang. I remember going out with him in about 1928, after he had his commercial licence. He was hoping to get two bucks out of me to take me up. I was dying to go, but even more I was desperate to hang onto my two bucks, which was the only two bucks I had. I kind of felt Rupert ought to find a way to get me up for free. He had the use of a plane that belonged to Sprott-Shaw Schools, a low-wing Barling, two-place. Considered quite a hot little aeroplane at the time. While we were standing around he introduced me to some of his pals. There was one girl who was apparently worth a lot of money and loved to fool around out at the airport with these pilots. I believe she even flew a plane herself. My friend Rex Chandler remembers that she belonged to a prominent family by the name of Mackenzie. She was driving a Cord with front-wheel drive. A huge great black thing with just two little seats. I was terribly impressed, and the thing that impressed me most was that it had *two* taillights! On one car! It was the first time I'd seen that. Then all of a sudden somebody said, "Hey, Rupe! Got a customer!" There was a guy at the counter so Rupert said okay, I'll take this guy up and you stick around here till I get back.

The customer sat in the forward cockpit and Rupert got in behind. The mechanic started the prop by hand and away they went. They said he'd be back in fifteen minutes. About an hour later, still no Rupert. About an hour and a half later he and his passenger turned up in a farmer's truck. He'd had an engine failure and made a forced landing on a farm towards Boundary Bay. He took the man up to the counter and made sure he got half his money back, which I thought was pretty decent, for Rupert.

That let me out. I didn't have to fly with Rupert and part with my two bucks. Less than a year later I had a photograph from Rupert showing him in his flying gear—his sort-of riding britches, leather leggings, his silk scarf wrapped around his neck and his flying helmet with the earflaps hanging down—looking very aviation-y. He and his pal Gordie Bolger, also dressed up, were standing side by side in a field near Vernon and between them was the Barling. The engine had come out of it, the undercarriage was collapsed and the fuselage was all twisted up, it was completely wrecked. Under the picture were the words, "Down on the Farm." That was all. The story, as I learned later, was that it was really rather underpowered and going up to Vernon he was clipping it a little close to the hills. He got into a downdraft and the thing wouldn't climb fast enough to

Cousin Rupert in the Barling trying to get me to chance my first flight — and part with $2.

The day I did take my first flight. Bush pilot Grubby Grubstrom barnstorming in Fairchild Razorback at Savary Island circa 1933.

View from my bedroom at Savary Island, showing my ham station antenna.

keep up with the mountain and he stuck it right in. He wasn't very popular with Sprott-Shaw Schools after that.

All of these adventures gave me a feeling of being connected with the flying world. I would have loved to go down and learn to fly, but knew I couldn't afford to, and I envied Rupert very much.

My father and mother built up a fine modern farm on the family property in Whonnock, but in 1912 they sold it so they could build a yacht and cruise up the coast to Savary Island, which they had fallen in love with while honeymooning there in 1898. The family who bought the farm, Watson by name, defaulted on the payments, and then when the war started in 1914, took refuge under the wartime moratorium on debts. The result was that we were left living in a tent on the back of our old property. In 1913, we moved to Savary, but not in a yacht. My dad traded work for a free cabin over the winter, then in the spring we found an unused road right-of-way and pitched our tent again. We squatted there in varying degrees of poverty for ten years. My mother was an outdoorswoman and loved it, but it was quite a comedown for my dad, who had grown up in a privileged family in Derbyshire and had studied medicine at Cambridge.

Savary had a one-room school which went up to what was called "Fourth Reader" or high-school entrance, and I slogged through that for the first four years we lived on the island. I "graduated" in 1918. There was nothing more I could do in the way of educating myself, and my mother had me taken on as an apprentice officer in the merchant marine. The deal was no pay the first two years, fifteen dollars a month the third year and twenty-five dollars a month the fourth year, by which time I would receive my second mate's papers. I landed up on the *Melville Dollar* with Canadian officers, Scotch and Welsh engineers and Chinese crew. As we wallowed across the Pacific I immediately became seasick, and I was still seasick when we hit Japan. Most of the ships we passed were still in camouflage paint and there were still many square riggers, many schooners, five-masted schooners, all carrying lumber. The *Melville Dollar* was a steam packet, two well-decks, ten thousand tons, but the experience of the officers was, as they say, "before the mast." I wasn't permitted to talk to an officer except to say "Yes, sir!" and the Chinese crew couldn't speak English, so it was a pretty drab affair. So drab, in fact, that after about six months of it I'd had enough and my family had to buy me out. That was as close as I ever did come to having my second mate's papers, but in a roundabout way my stint at sea did end up steering me into a career.

There was one person aboard I found I could talk to and this was the radio operator, who was from Courtenay. He befriended me, and when the skipper wasn't around I could go up to his cabin and listen to him send messages and listen to the messages come in. I became very interested in the ship's wireless, which was a reasonably new one. I got a job in the woods when I got back to Savary, ran a logging donkey for a while, got my steam ticket, but my plan was to save enough money to buy some parts and build a wireless set. Isolation was something all of us felt very strongly living on Savary, and the prospect of being able to break through the long winter silence with wireless was very exciting to me. I saw it from the start as something that had special application to all those of us living in the remoter places along the BC coast, cut off from the world but still intensely interested in it. I learned Morse Code and made a little receiver that I could use to listen in on ship-to-shore transmissions in code. Eventually a few radios started to come around then, about 1922, the very first crystal sets—you could hear voice and music from KPO San Francisco and CKCD Vancouver on a cat's whisker and a good long piece of wire, and I made some of these. And then I bought some of the early tubes and made a set that worked rather well, and first thing I knew people wanted me to make sets for them. Some bought sets ready-made from the mail order houses, but they didn't work up the coast so they'd bring them into me and I'd try to improve them. So I quit logging and went into radio work full time, on my own. I couldn't earn near as much, but from the start I was anxious to be in business on my own.

As much as I could, I avoided the other type of work, which was digging wells or cutting wood or doing anything you could get to do—this was into the Depression now, pretty tough times. I could repair all the sets that came by, or rebuild them—I used to take a lot of straight TRF sets and make them into superheterodynes and that made them work like anything. So up and down the coast I began to get a bit of a name. I put up a sign in the Ragged Islands about ten feet square: "Radio Expert A.J. Spilsbury Savary Island." The result was that some of the tugs started showing up. They'd leave their booms on the other side and steam over: "Where's this radio expert? Our radio don't work!" This is the way it started.

Dr. Lea, one of the summer residents at Savary, took me aside one day and said, "You know, Jim, I hate to say it, but you're wasting your time by staying up here. Down in the city we're desperate for people who know radio. With your knowledge you could step into a real career." But I was a country boy, and the thought of moving to the city, where I would be up against all those thousands of better-educated, better-dressed, better-off city guys who knew the ropes, filled me with panic. Here up the coast I had things to myself. I understood the coast and felt its deep need for what radio had to offer. All these little camps and homesteads hidden away in bays and inlets all the way to Alaska, cut off from the world, waiting for someone to come along and plug their cookshack into the Metropolitan Opera. I suppose I thought there were more of these people than there probably were, but in those days there *were* a lot, several thousand perhaps, as opposed to only several hundred now. It seemed a big enough field for my talents; all I had to do was figure out how to get out there and reach them.

The Radio Business

THIS WAS WHERE I CAME UP AGAINST the second cause of isolation on the BC coast, after lack of communication: lack of access. Everything had to go by boat. In the very first days of exploration, the sea served settlers well enough, allowing them to penetrate up every inlet and to the farthest tip of Vancouver Island and the Queen Charlottes. But the sea is a slow and unreliable highway, and as settlements grew and farms developed, the pioneers found they were just as isolated as when they first came. No roads or railways grew up to take their produce to market or bring new settlers out. Every time the government thought about building one, their engineers were stopped by the sheer cliffs and deep inlets that make it the place it is. A steamer service grew up to serve larger stops like Pender Harbour and Alert Bay, but it was limited. It brought loggers north and took expectant mothers south—if they planned things very carefully—but it was no good for taking a doctor around to his patients or a radio man around to his customers, because it didn't stop everywhere. It was very clumsy. Most of these people you could only reach by taking a lot of time and going around in a small boat. For city businesses and services, that was too much of an ordeal, but I had grown up on the water and I had been in and out of small boats as long as I could remember.

On Savary everyone had their own little boat for personal use and we kids grew up not riding bikes, but instead, banging around in rowboats. As a boy gained in years, he dreamed not of owning a jalopy, but rather of getting a "kicker" boat with an engine of some sort.

We couldn't afford any sort of boat at first, but on Savary Island you could find almost anything on the beach if you were prepared to wait long enough. One stormy night in 1922, a Polish hermit wrecked his small boat on the south side of the island, but when my father and I went to collect his things the next day the old man forbade me to salvage the little Dunn marine engine that had brought him to grief, crying "Leave her lay where Jesus flang her!" After the old man had been gone long enough for the apocalyptic effect of his curse to wear off, I went back with a cross-cut saw to cut the log away that was on top of the little motor, and managed to get it home, where I worked on it all winter and finally got it to run. That same year we found another old boat on the south shore. It was a heavily constructed work-boat about twelve feet long, and had probably been lost from the construction site of the new docks being built at Powell River. Aside from one large hole in the bottom it was in sound condition and I was able to repair it and install the old two-horsepower Dunn in it. I was delighted with it, but it was much too heavy to haul up on the beach every day, so it had to be anchored out, with all the attendant problems of a small open boat and Savary Island storms. Then one day, not long after launching it, I received a cash offer. Norman Palmer, then twelve and the youngest son of Al Palmer, boss logger of Palmer-Owen Logging Company, saw it one day in Lund and just had to have it. He was making twenty-five dollars a week working in camp as a "whistle-punk," and could easily afford the total fifty-dollar purchase price I was holding out for. It was this money I used to buy the parts for my first experimental crystal set.

Another boat drifted into my life in a different way. Mother's old friend Ethel Burpee shared Mother's passion for the active life, and she got to spending a lot of time with us. She had some money from working as a school teacher and bought a very well-made twelve-foot rowboat of rather distinctive design, originally built in England. It was light enough to pull up the beach and much more practical to keep. She donated it to the cause and I horsed around with it for years. Dad helped me rig a mast and lug sail for it and I learned to sail. Then, in 1914, Dad bought one of the first outboard motors ever built.

This was the early model Evinrude, that cost $125. It had one cylinder, developed 1½ horsepower and weighed seventy-six pounds. To start it, you pulled the black knob on the top of the flywheel. Then when you turned around and grasped the tiller to steer it, you invariably received a sharp crack on the point of the right elbow from the wooden knob, which, if the engine was turning up maximum revs (600 rpm), gave you quite a nasty jolt. I have a permanent chip out of my right elbow. It was usually very difficult to start, sometimes requiring fifteen minutes or more of frantic jerking on the knob. The knob was about two inches high, so that the entire strain came on the fourth finger of your right hand, which usually became semi-paralyzed after the first half hour or so.

After a year or so of this difficulty we discovered, quite by accident, that if you reversed the process and cranked it backwards, it would start willingly with the first or second pull, and would run much faster in this direction. We ran many miles, steering backwards, to the huge delight and derisive cat-calls of the other rowboats, which didn't have this new-fangled device. Usually, though, after running backwards for about a hundred yards, it would agree to being started in the forward direction. I had that engine for ten years, and it was only after selling it to Tom Lea for twenty-five dollars that the problem was explained. Apparently, at the factory, the piston had been installed upside down and, being a two-stroke engine, it didn't suck the mixture in from the carburetor properly due to the offset shape of the piston head. Tom was the first one to take it all apart and find this out.

There was another feature of the design of this engine that was something less than perfect. The engine clamped onto the transom of the boat by means of a cast-iron bracket and two thumbscrews. The driveshaft housing was a piece of 1½-inch brass pipe, two feet long, with the bevel gear housing and propeller bolted to the bottom end. The gears required stuffing with axle-grease at least once a day or they would rust out. There was, of course, no provision for tilting the engine up out of the water, so if you ran over a log or hit bottom it would either snap the cast-iron bracket in half or rip the transom out of the boat, but in either case, the engine would go overboard. After the first few times, we learned to tie a rope to it with the other end secured around a seat.

There was a third way to accomplish the same effect, and this always caught us by surprise. This one-cylinder engine, turning over at 600 rpm, created unbelievable vibration, which very soon loosened all the nails in the boat, causing it to leak, as well as giving most passengers a certain numbness of the posterior regions; but there was another more serious side effect. This vibration would loosen the thumbscrews on the transom bracket, and when you tried to execute a sharp turn to port or starboard the whole engine would tilt up sideways and the propeller would come right out of the water, sousing everyone within reach, before the whole caboodle went overboard. Different procedure, but the same net results. We still needed the rope around the seat.

Eventually, of course, newer and better outboards appeared on the market. They were lighter, quieter and much more reliable. But in spite of our first experience, Dad would not admit that there was anything wrong with the Evinrude line. He read and believed all the glowing advertisements that appeared in *Rod and Gun* and other magazines. In 1924 he bought one of the newly-announced and latest Evinrude products.

This was the Evinrude Big Twin. It was claimed to have broken all world speed records for outboard-powered craft, and cost four hundred dollars. It had two cylinders, developed 4½ horsepower and weighed 114 pounds. They had increased the rpm to an unbelievable twelve hundred revolutions. In all other respects it resembled the original engine, even using the same cylinder castings, and the same overall construction of cast-iron and brass pipe, with no tilt-up feature. Even though the new model had the two cylinders opposed, it vibrated twice as badly as the single. It was so bad, in fact, that the vibration cracked the mounting bracket the first week we used it, and we had to take it over to Frank Osborne's Lund Machine Shop and get it reinforced and welded. Then the flywheel came loose from the hub and shaft, and shot, spinning like a top, into the bottom of the boat. We took that into the machine shop and had it re-riveted to the hub. Then the screws in the bottom end of the brass shaft worked loose and the whole bottom end dropped off in deep water. This all happened in the first three months. The Evinrude company was no help in any of this. That, I think, was about the time they closed their doors and merged with Johnson Outboards, and Ole Evinrude himself started his new company, ELTO (Evinrude Light Twin Outboard).

I forgot to mention that the magneto, which was in the flywheel, had conked out quite early on and I had modified it to use a Ford spark coil and dry batteries. The rebuilt transom bracket worked for a short time, but resulted in transferring all the vibration stress to the lugs on the engine base casting, which in turn tore away from the base, once again leaving the engine hanging on the rope. Now, with no bracket, and no bottom end or propeller, I resorted to drastic action. I made up a kind of cradle to mount the engine on its side and sold it to an Indian from Squirrel Cove as an inboard for his canoe, complete with spark coil and dry batteries. It was still running when he passed from sight on his way to Squirrel Cove.

We went on to other kickers, but nothing of this order was going to help me get up the coast to sell my radio services. As time passed I became more excited with the idea the radio business of the coast was all mine, if only I could get a gasboat large enough to take me around, but gasboats were so far out of my price range I couldn't even think about it. But I kept looking.

Finally I made a deal with an old Swede in Lund named Eric Nelson, who had an equally old codfishing boat, the *Mary*, with a nine-horsepower Buffalo motor. The engine stood about four feet off the engine bed and you started it with a Johnson bar. Punk punk punk it went. I plugged up and down the coast with this, called in at logging camps and homesteads among the northern Gulf Islands as far up as Loughborough Inlet. I didn't carry very much with me. I had a minimum of tools, a few spare tubes, a few 45-volt B batteries, 22½-volt C batteries, a few spare transformers and so on.

I made my first trip in the *Mary* in April, 1935. I worked out my average net take, after deducting grub, fuel and a dollar a day boat rental. It came to $2.63 per day. It wasn't a quarter what I could make logging, but the fact it was "profit" rather than "wages" made it ever so much more precious to me. I was just *thrilled* to be in business for myself. Grandfather Spilsbury, the head of the Spilsbury clan back in Derbyshire, had had the honor of naming me, and he had chosen to call me after the man who had established the family fortune digging canals in the eighteenth century, James Ward. The rest of the family were pretty much a bunch of genteel do-nothings, and the Governor, as everyone called my grandfather, had nearly run out of money by the time I showed up in 1905.

"Name him after the only member of this family who ever did anything," the old boy grumbled. "It's high time someone else in this family made some money!" So I was named Ashton James Ward, and from the time I was conscious of anything, my family was telling me it was my destiny to get on in business. I guess it got lodged pretty deep down inside, because when I finally got the chance I never hesitated. It felt just wonderful, as it must feel for the would-be painter who gets his first show, even though he gets less for it than if he spent the time picking rags. On the *Mary's* third trip I managed to increase profits a hundred percent to $5.84, and by the sixth trip in May–June 1936, I had it up to $7.03 per day. Mind you,

My grandfather, the Rev. B.W. Spilsbury, and his wife, about the time he ordained that I should get on in business.

Charter members of the Island Net: Jack Tindall (VE 5MK), "Hep" Hepburn (VE5 HP) and myself (VE 5BR.)

Our floating home and radio shop, the Five B.R., *docked at Lund.*

this only lasted forty days of the year, then it was back to chopping wood and digging wells.

Things were going so well I decided to spend five hundred dollars refurbishing the *Mary*, but I had only enjoyed it for about one trip when old Eric Nelson took it back and sold it to someone else. At this point the Island Net came to my rescue. I failed to mention that in 1926 I had passed my code test and got my amateur radio transmitting licence so I could set up as a ham operator. I was NC 5BR, later VE 5BR, and now since they've changed the district arrangement in Canada, VE 7BR. This was the real breakthrough as far as my own isolation was concerned, and I made a lot of friends over the air. Together with a bunch of other hams along the coast we formed a regular group we referred to among ourselves as the "Island Net." One of the regulars was Bob Weld, a fine old gentleman in Parksville. He and his son Brian had spent two or three years building a pleasure boat, but Brian left to become radio operator for the BC Police in Victoria and the old man found the boat was too much, so he said one day why didn't I buy it. It was forty feet long, very beamy, and had an old Cletrac engine salvaged out of a tractor. It was the answer to my prayers, but I said, well, gee I'd love to but I haven't got that kind of money. Well, he said, you were paying Eric Nelson something—you must have some money. I said no, I was just paying him one dollar a day. Well okay, he said, I want twenty-five hundred dollars and you can have twenty-five hundred days to pay it. So in October of 1936 I purchased this boat and registered it in Powell River as the *Five B.R.*, after my ham call. Powell River had just become a port of registry and this was the first boat ever registered there.

It was a wonderful boat. Very commodious. I had a stateroom forward with two bunks, a roomy wheelhouse, a seat that made down into a double bunk in the main saloon and behind that a small room as a shop. I could go seven miles an hour, which I thought was marvellous. It increased my range greatly, and I began travelling further up the coast, eventually as far as Seymour Inlet and the north end of Vancouver Island. I spent the next seven years on that boat. I was married in 1937 and lived aboard with my wife Glenys; my eldest son Ronnie, who was born in 1940, spent his first two years aboard the *Five B.R.* It was our floating home.

When the war started up I tried to enlist in the Air Force, but they took one look at my diploma from the Savary Island School and gave me the bum's rush. At that stage they were only taking university men. So I concentrated on my radio work. Radio-telephones were starting to come in.

The first radiophones on the coast had come in around 1921 or 1922, supplied by the Canadian Marconi Company. They were very large, very old-fashioned of course, with two enormous tubes and two huge helixes made out of copper ribbon. Only a government forestry launch had room to put the darned things—they were as big as a refrigerator. The Marconi company wouldn't sell them, they would only rent them for sixty dollars a month, and they had a darned good thing going. They said they had a monopoly. Then a chap named Chisholm started building equipment for the forestry service and there was a lawsuit, but he came out of it alright. Along about that time I started to get into it in a very small way myself. I was starting to build them on my own—the first one I made went into Theodosia Arm for the big Merrill, Ring and Moore camp. Later I built more for the Provincial Police and various camps up the coast. But between running the *Five B.R.*, repairing household sets and doing everything else, I was having a hard time keeping up, as I explained in a letter to one of my ham friends.

My first boat trip netted me $2.63 a day and it made me feel virtually independent. Now the net earnings run about $100 a month and I feel almost wealthy. The increase in earnings up to that level was quite steep and then flattened off about 1937. The reason for this flattening off is quite obvious to me. Since 1937 I have had more work than I have time to do. New business has exceeded my capacity for work. The last year has brought a certain amount of new radio-telephone business—about three months of the last twelve have been devoted to this—and the time needed for this has been at the expense of my regular household radio customers so the net gain was practically nil. A proportion of my household clientele is mad because I have neglected them. I don't know what 1941 will bring, there is every prospect of an increase in the radio-telephone work. On the other hand, any further neglect of my household clientele is likely to be disastrous.

I wasn't making up the disastrous part. I had a letter on my desk from one Francis Dickie of Heriot Bay, or as he was invariably called around those parts, Francis Dickie *the writer*. He was by his own admission a top author, a familiar of Somerset Maugham and creator of such sadly underrated works as the novel *The Master Breed*, so his letter may be of historical interest:

> GENTLEMEN: Since nearly a year ago a bill has been upon my file, for the sum of $2.65. This would have been sent long ago except for the fact that, since it was *then* two years since last you called here, I naturally kept expecting you in at any time, and let the bill go to pay you when you came. I particularly expected you because over a year previously I had written you the new machine which I purchased from you needed going over. I would have sent the machine down, only for two reasons; expecting you in at any time; and because the box belonging to it never yet has been delivered. This you promised to do. You evidently have forgotten, but when you sold me this machine there was no box, and you said you would bring it along the next time you came by.
>
> It is now two and a half years since you called here. Many times you have been very close. Last autumn when you were at Redonda Bay, every machine in the place was out of order at the same time. As you were so close and had not been here for so long, we were all expecting you. You have been right on this very island; and yet you passed this place. If you have decided never to call here again, it would be at least something to tell me, and then one would know what to do when their machine goes out of order.
>
> For six years I was your customer, buying everything possible from only you; this by way of appreciating your regular calling in here—part of that "enthusiastic support" you used to speak of in your circular letters.
>
> In good Canadian vernacular; I have never experienced such lousy treatment.
>
> Francis Dickie

There was no reason Dickie should have had trouble using Canadian vernacular, because for all that he tried to affect the manners of the displaced Englishmen who were all around him in those days, he was born in Saskatoon. I had some reasons of my own for coming down the other side of the channel when I went past Heriot Bay. Dickie was always kicking for service but never paying when he got it. The $2.65 was for a tube I shipped him about two years earlier, but I knew if I went in he would pay me for the tube and want a twenty-dollar set of B batteries on the cuff. Still, his loud cries of anguish were not doing me any good, and there were others equally neglected whose business I did want to keep.

Here again the Island Net came to my rescue. I was able to arrange with another ham friend to take over the time-consuming work of building my radio-telephone sets. This was a younger man named Jim Hepburn, who at that time was working as office boy and resident radio expert for Island Tug and Barge in Victoria.

He was my match in terms of radio theory, and the building he did was very sound, although he hadn't as much practical experience as I and tended to leave loose ends. He was a bit of an odd bird, the kind of character who wore thick glasses and stayed in his room doing experiments while the other guys went out with the girls. (I did a lot of experimenting in my room too, but I had an excuse—for ten months of the year there were no girls on Savary!) He was rather frail and always seemed to have some sort of ailment to tell you about, and when he tried to join up he failed his army medical. But there was nothing wrong with his mind, unless it was a slight devious twist. People would remember him years later as a shadowy figure drifting around the office in crepe-soled sneakers reading their letters upside down. I didn't know this side of him at first, and we struck up an enthusiastic correspondence full of ham lingo.

KEY TO LINGO

73	best regards
1776	the radio frequency our net used
BL	Bob Weld, our net friend in Parksville whose call was VE 5BL
xtal	crystal, as used in radiophone tuning
FB	fine business
Hi!	laugh.
MIM	exclamation point

MK	Jack Tindall, a close friend of mine and net member in Refuge Cove whose call was VE 5MK
NM	no more
OM	Old Man
tmw	tomorrow
VAB	Vancouver
VAK	Victoria
yda	yesterday
YL	young lady
MSG	message

Dear Hep:

It was a very nice surprise to receive your letter yda on our return after a four weeks trip through the north part of my district. I feel all pepped up again just like I'd had a good long session with the island net. Boy how we miss the gang on 1776 MIM. Although I have not been able to take a very active part in the nightly confabs since living on the boat it is still very comforting to be able to listen in and hear the familiar voices. [Later in the war hams were banned and not permitted back on the air until November, 1945.] I will be seeing MK this coming week, and after he has had a gander at your letter I will forward it to BL and make it an excuse to write the old boy.

Being securely married and nearly a parent, I adopted a fatherly tone in personal matters:

Glad to hear you are showing a little interest in YL's and things now. I always figured you were a little lacking along those lines and rather admired you for it. Don't forget this business can be easily overdone however, especially when the balancing influence of ham radio is not present. Hi!

Things went so well with Hep I was soon proposing he join the business. No small part of this was a simple desire to have someone to talk to about the intricacies of marine radio, but I began working out more serious reasons. When I was up the coast in the boat, people would write in to Savary and say would I please come to Telegraph Cove or Pender Harbour or Roy or Seaford, they had a radio to be fixed or they were thinking of getting one of these new radios you could talk into. I had pre-printed post cards I left around for these people to send in when they needed me, and they'd put them on a down boat (southbound steamer), the boat would put them off in Vancouver, eventually an up boat (northbound steamer) would take them back up to Savary, and later yet I'd come by Savary to get them. People would go months without their radios, waiting for me, and by the time I finally drifted in on the tide to their camp at the head of Bute Inlet or wherever, I'd find they'd long since taken care of their problem some other way, generally by shipping it to Vancouver. Only in the rarest cases would they have sufficient confidence in the Union Steamship Company to send the faulty set to me at Savary. As the radio-telephone business developed, I found I was missing out on a lot of big sales, not because people didn't want me, but because I was up the coast chasing my tail somewhere when the opportunity arose. I reasoned that the best solution would be to establish a mailing address in Vancouver with someone to man it and broker the work for me. If that person was Hep, he could also have a shop and fill up his time building radio-telephone sets. Since I was paying him to do this anyway, the difference in my outlay wouldn't be that great. As I wrote Hep in November, 1940:

> The increased efficiency offered my upcoast business by having a Vancouver connection, and the added prestige of a Vancouver address on the stationery will I think have a cumulative effect. I believe it will only be a short time before you are kept as busy as you would want, and I think the Vancouver end, due to a much lower proportion of non-productive time steering the boat, copper painting the bottom, fixing the engine, doing office work, running between ports, etc., will actually and eventually show a better margin of profit than the boat, although the boat remains a sort of necessary evil to maintain our coastwise contacts.

If I'd had an inkling just how true those words would prove, and in just how short a time, I'm sure I would have been quite dismayed. Hep wanted to get into the business like nothing else, but being more than a little cagey, he let me think he was having a hard time making up his mind.

> The more I think of your offer the better it sounds. The only part I don't like is the absolutely final uprooting from the green fields of VAK to the foul backwater of VAB. You are more or less based on your old home at Savary whereas it will cost me a good deal of money to go see my folks even occasionally. The YL don't think too much of the idea on that account. Hi! If I stayed on here I could reasonably expect to marry the gal in about six months and the Vancouver move would put that off a year at least. That is the big sticker OM, we have been fixing up our little shack at Brentwood since it was definitely seen that I wasn't going to get into the service and altho no real plans have been made it's just a matter of holding off til the old wad is thick enough to get started without too much skimping. Those are the only two objections I have to the move and as you may see they are personal and no doubt selfish. Hi! So we might as well go on with the discussions, and see if we can work out the details so that you don't go broke over the big expansion and I don't find myself giving up a nice soft job to go tearing around Vancouver for nothing flat. MIM

I had started, I think, by offering him fifty dollars a month plus half of any profit the operation earned over a thousand dollars, but in the end we settled on ninety dollars a month, turned my business over to a limited company, and gave Hepburn forty percent of the shares as part of his monthly pay. I had pushed for keeping the firm in my name, saying, "I hate to think of having A.J.Spilsbury Ltd. splattered across the countryside, tacked to stumps and emblazoned on the pages of coastwise trade papers etc. but am thinking it is expecting almost too much to hope that the average up-coast customer is going to recognize any connection between a Vancouver Co with a strange new name like Marine Radio Technicians, and the old 'Jim Spilsbury Fixem' they are used to, and forthwith entrust the innards of their pet radio to the tender mercies of a new bunch of town slickers." But Hep wasn't about to have his name left out, and when he proposed "Spilsbury and Hepburn Ltd." I couldn't think of any reason against it.

I borrowed fifteen hundred dollars from my wife, who had come into a small inheritance, and we built a tiny little box of a building down at the foot of Cardero Street, which is still there. Hep got set up there doing inside work and we could send messages back and forth.

Mostly we sent letters, cursing the erratic mail services of the Union Steamships as we went:

Lund, April 22, 1941

Dear Hep:

So far so good. Arrived Savary this afternoon, got blown out with a westerly at suppertime, now holed up in Finn Bay for the night. Will check mail tmw morning then proceed to Twin Islands and if I can get through there beat it for Refuge Cove tomorrow morning.

On the way up from Vancouver last Sunday received a msg from Lazo [the radio-telephone exchange at Cape Lazo] to contact the *Pacific Foam* off Trail Islands and towed a couple miles backwards with him while I gave the set a going over. Don't know what the kick was because everything checked out FB. Will charge Doug Coyle [owner of Pacific Coyle Navigation Co. Ltd.] for a service call anyway and the skipper gave us a couple salmon for good measure. MIM Cleaned up a few jobs around Pender Harbour and ran an ad in the local paper announcing new shop etc. Sure will be glad when the new stationery arrives as am getting tired of writing in the new address on everything and trying to explain the setup.

Well, that's the works for now—

Dear Jim:

Have to report that I have been off the job for the last few days with my old friend stummick fever. Still feel pretty wobbly but got down again this morning and found the box simply bulging with mail.

Spilsbury and Hepburn head office, 1941.

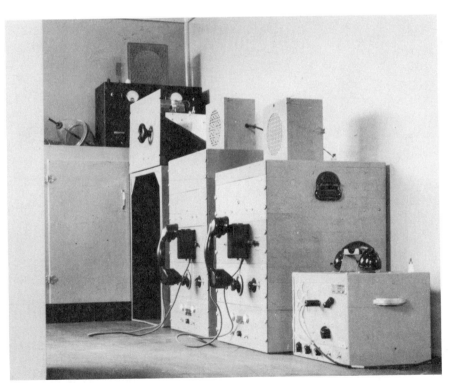

Early S & H radiophones.

Glenys and Ronnie at "the roaring hole," Acteon Sound.

The fanciest floathouse on the coast. Visiting Maude and Oscar Johnson, Seymour Inlet.

Went into the printers this morning and collected all the printing and am sending you some stationery, envelopes, sales books, tubestickers, half the address labels and 1000 post cards at Skookum.

—Hep

En Route Chonat Bay, BC

Dear Hep—

Pounding this out while the wife steers a bumpy course through a SE chop and veiled moonlight. Looks like a storm coming up and I want to make Chonat Bay tonight so I can chuck this on the steamer tomorrow. Received the confusion of mail and parcels and freight generously mingled with gravel and seaweed and caulk-boot splinters. Thanks for your promptness and I hope you're feeling better by now.

Worked Lazo today and you coulda brushed my eyes off with a feather when the operator comes out with, "5HP and 5BR, oughta make a pretty good combination good luck OM." It was Glen [an old Island Net member], back there relieving for a few days. Guess Lazo must be wired up to the grapevine like the rest of this durned coast. When we came up from town I go hiking into 5MK to let him in on the news [of the new partnership]. He ups with, "Oh yes I heard about that some time ago!" Guess I gotta have a faster boat if I'm going to spread any of my own gossip in future, along this coast.

I'm in a good humour. Sold my second Stromberg-Carlson in two days and got full price for it. Picked up four while in town at $30 ea. MIM.

Chonat Bay

Dear Hep:

We have just arrived here from Rock Bay. The camp is in darkness so I guess I don't go ashore til morning.

Darned if I didn't get kicked out of a camp yesterday. First time I've had that happen. Camp "O" in Nodales Channel is under new management. I called there yesterday as usual and went ashore to see if there was any business. A guy met me on the boom and led me to the office. I walk in all smiles figgering the boss must want me to fix his set. To my amazement they asked me very politely to get going. After awhile I got them to loosen up a bit and explain things. Apparently our friend Richardson of the *Ida T* [Acme Radio Service] had called in a coupla months ago and took them for a ride. He sold them three machines in the bunkhouses all of which promptly went on the bum a few days later. There was such a stink the OM says NM that sorta thing.

Barney Bendickson, captain and owner of the *Sea Spray I*, a new small tug is in line for radiophone according to his brother Arthur Bendickson at Boughey Bay. At his suggestion I am writing him as otherwise he is bound to fall into the clutches of Canadian Marconi Co. Arthur Bendickson, although a small camp, is a very progressive sort of duck and is a decided prospect himself.

Called at Western Logging Co. Forward Harbour. Local boss is Mr. Norman. Seems definitely interested but has been having some tough luck lately. Just broke the second crankshaft in their new diesel yarder and don't feel like asking owner for money until the logs start to roll again.

Called at Topaz Timber Co., Topaz Harbour. Walked over the hill and called on all houses but no work. They had expected us last Xmas but heard we were both drafted into the army and business closed up. Same at Eng's camp. Wonder if this is another of Richardson's fables?

Find that Eng's camp is now very interested in phone. They have been going across using the Topaz rig and realize its usefulness. Eng is another small operator and generally considered very backward and slightly haywire, but never seems to go broke and is an oldtimer.

Made a rather good contact at the fish scow up Loughborough. Kenneth Brown the buyer says he sent at least a dozen radios down to Spencer's [a

Vancouver Department Store] last year and there wasn't a single one of them ran right after. He had obtained a copy of our circular letter and had it nailed to the wall intending to send the next one to us. I encouraged him no end. Anything Brown sends down will be care of Francis Millerd Co. and they will pay. Brown himself I would not trust with a nickel. Also I would be quite careful with Millerd Co. They have always paid me OK in the past, but they don't have a very good reputation.

We ran into the notorious maverick cannery mogul Francis Millerd again at Redonda Bay, where he had re-started an ancient cannery that hadn't run since the 1920s:

Cannery at Redonda Bay going full blast now on herring and probably will keep running for several months to come. As long as it lasts will have to make frequent calls as the workers sure can't hang onto their money. Not that I enjoy the place very much. We have to tie up alongside the offal scow at the wharf. All the splits, culls, rejects, heads, tails and gizzards are dumped in a heap on this scow. Every three or four days when they figure it's about due to capsize, they tow it out into the bay and dump it. In the meantime these fish started out on the west coast or somewhere and by this time getting a little over-ripe. Then when the boys toss em down onto the scow their marksmanship is not the best, specially at low tide with square-nose shovels, and what doesn't make the scow either splatters the *Five B.R.* or goes overboard, sinks to the bottom and lays there in a layer about two feet thick waiting to be exposed when the tide goes out. To add to the normal wastage, they accidentally dumped 16 tons off a scow a month or so ago and boy! You should just sample it some night when the tide is low with an offshore breeze. Then there is a thick scum on the water, washings from the cannery, consisting of herring fat, slime and scales etc. that cakes itself all over the wharf piling and particularly the overside ladders so that everyone as a matter of course wears gum boots and goes around with an inch of the stuff sticking to them. It tracks into the boat worse than sticky snow and by the time half a dozen customers have been in to look over my stock it's a mess. Just to top it off there's a colony of seagulls numbering in the thousands that share with the *Five B.R.* the joys and conveniences of the offal scow. They take it turn about, as the scow won't hold them all at one time and half of them feed while the other half circle around overhead and watch the fun. I have spent all afternoon at Stuart Island with the hose and scrubbing brush working on the upperworks, with half the population in town looking on and asking where the heck we've been anyway.

Apropos of this, Redonda Bay is ripe for radiophone. The foreman, Henshaw, makes three or four trips a week to Refuge Cove to use Jack's phone. Both Henshaw and the cannery bookkeeper say Millerd is due to snap at any minute now. So is Jerry Olmstead, the logger across the bay, who has been talking about putting a phone in his boat the *Roanoke*. I think they're both stalling waiting for the other to do something first.

And so it went. From the first, Hep was completely snowed under at the shop. We both worked incredibly hard for incredibly tiny sums of money. There were times when neither of us took our salary because the bank account wouldn't stand it. I chased all over the coast trying to collect bills of twenty or thirty dollars. My nervousness about the new company's climbing expenses made me overscrupulous about details: twice I wrote Hep hounding him about using printed envelopes to write me when plain manila would do. Finally I purchased a bundle of the plain myself and sent them to him. In October I was after him about charging Milton Lilja of Read Island $3.50 for rewinding an oscillator coil, saying "I always get $4.50 for this job."

We kept a wary eye on our competition, relishing the supposed follies of Bob Richardson, whose radio boat was called the *Ida T*. "Does he tie up at the Eccles' dock when he's in Vancouver?" Hep wrote. "I don't know where the *Ida T* ties, but wherever it is, it won't be far from a beer parlour!" I replied. Every tube and table radio sold by the Canadian Marconi Company brought us pain, and we seldom crossed their path without bringing back a story about how clumsy, outdated and overpriced "Macaroni" equipment was. Even when Stuart Island machinist Jack Parrish got a ham friend to build him a homemade radiophone, we were full of snide comments about the "homebrew heap," and I wrote Hep that we were lucky to be rid of Jack's business "as he is an inveterate kicker anyway."

But the wartime economy of the coast was heating up and business kept building in spite of itself. Sales in April 1941 were $428.92, in May $1,023.43, in June $1,271.43. Even with Hep building every spare minute we couldn't make enough radio phones to keep up to demand so in July we hired our first employee, a relative of one of Glenys' friends, named George Taylor. We got him under the apprenticeship program for seven dollars a week, leaving Hep fretting, "I sure hope I did the right thing!" At my urging Hep was taking only fifty dollars of his salary in cash, half what he had been earning at Island Tug and Barge, which raised hell with his wedding plans. He kept making jokes about his YL wanting to "toss a brick in my bilge" but got married in October regardless. I was in Pender Harbour and didn't come in for the wedding but added a short congratulatory paragraph to my regular parts letter. "Marriage is like bathing at Savary," I wrote him, "the look of it scares you, the worst moment is just before you get ducked, but once you're in it's not too bad."

In November Hep brought the new wife, Molly, up to see us in Lund. We liked her, and she had given up wanting to put a brick in our bilge. She was quite intrigued with all our doings and took a great interest in the office, standing in at the phone during Hep's sick spells and service trips. She added a pleasant touch to our company relations, rushing around at the last minute to get fresh bakery bread to put in our parts shipments and sending along surprise gifts for Ronnie and bits of gossip from the city.

In 1942 the huge Wood and English holdings at Englewood on Vancouver Island were taken over by a group which would become Canadian Forest Products, although it was not known as that at the time. Our good customers Brown and Kirkland were involved in it and we got the job of installing not only two sizeable radio-telephone stations, but most of the camp wiring as well. When I discovered I needed to make a quick trip back to town to do some alterations to the set I caught a ride south on the Union steamer *Catala*, leaving Glenys and Ronnie up there on the *Five B.R.* Then, when I missed the connecting boat back, old man Kirkland bought me a ticket on the Canadian Pacific Airlines Rapide, and I found myself aloft for only the second time in my life, flying over my familiar route up the Inside Passage. I told the pilot, Tom Lowrey, to land right beside the boat, planning to amaze Glenys and Ronnie, but it didn't work, as I told Molly:

> The trip back was lovely thanks, and well worth the paltry $28.75 that B&K are paying for it. Took just an hour and thirty-five minutes which is considerably under the time taken by the *Catala!* Glenys didn't appear in the least surprised although we landed almost on top of the *Five B.R.* Apparently concluded it would be me as soon as the plane hove in sight! If ever I put anything over on that gal I'll let you know. However she *did* seem pleased to see me, unlike Ronnie, who chose to ignore me completely for several hours.

I wrote everybody I knew about the flying experience, telling Hep in a separate letter, "The trip back was very nice. Took 1 hour 35 mins. and Englewood was the first stop. The dial said 110 MPH, the temp. 50, and the height I judged to be about 2500 most of the trip. Above Campbell River we went cross country behind some mountains on the island and dropped down on the straits just below Englewood. Very interesting to get a bird's eye view of some of the country I know so well from sea level. Had a pretty full load and they skidded down the river darned near to the mouth before he got her up. Passed over Savary at 40 minutes." I revised the details slightly for Mother and Dad, writing, "In order to save time I returned to Englewood Monday morning by plane. Lovely trip and most interesting to study the country from the air. All the coastline I know so well, and lots I'd never seen before. Passed over Savary forty minutes after leaving town. Unfortunately we were on the south side and saw nothing of you although we went almost over Indian Point. Landed right in front of the *Five B.R.* and gave Glenys a surprise as she had not been expecting me for two days."

Spending as much time as I did crawling from camp to lonely camp at seven miles an hour, I couldn't get over what a small place the coast suddenly became from the window of an aeroplane. I kept thinking about it and shaking my head once we were back in the boat taking a week to cover the same ground I'd flown over in that hour-and-a-half. It made me think a little differently about what aeroplanes could do on the coast, but they still seemed far from practical from an everyday standpoint.

Every now and then I'd get all excited about some new application of radio that might lead us to our financial reward by a more direct route, and we'd launch flocks of letters trying to work out details. Passing through Stuart Island in June, "I was tackled by a group of codfishermen about the possibilities of making up some device that will indicate the bottom depth for them. They say that if they can get something of that nature they are all in line to make themselves a fortune in cod." A device to measure water depth was the stuff of idle rumour then, and for months Hep and I fired off drawings of indicators rotating on old windup record turntables and sensors dangling overboard on lines. We finally wound down, but when I look back over our sketches I realize we were not so far from a workable version of the now ubiquitous marine depth sounder.

Stopping at Frank Kuchinka's camp in St. Vincent Bay in October, Frank hit me up about "a wireless signalling system in the woods." As I explained the problem to Hep:

> You are probably acquainted with the general setup used in the woods at present. A hooktender with several chokermen struggle around with the rigging amid a swelter of thick brush, mud, devil's club, splinters, caulk boots, thick gloves, foul language and general corruption, while a whistle punk (pronounced *vissel-bunk*) is perched on a stump at a safe distance, two or three hundred feet away with the whistle cord (if electric) or the more usual jerk-wire (if steam) in his hand. As soon as the big boys get the tackle on the log they scramble out of the mess and when the last man is clear the hooker lets out a blood-curdling yelp, the punk hauls on the wire and away go the logs with a terrific rending and crashing of fallen timber, rolling rocks and flying Swedes, etc. In five minutes the lines come back and the whole process is repeated, but each time they move on a little further so that every so often the

vissel bunk has to move on too to keep in earshot, and also move his *vissel vire*. This necessitates reeling the whole thing in to the donkey and starting out again as they log in a radial pattern like the spokes of a wheel. Usually one of the chokermen has to help the punk run his line out again and this takes considerable time. Then all sorts of little problems crop up such as trees falling over the wire etc. so the whole thing is pretty clumsy. On top of all this it means keeping one man doing nothing but sit on a stump getting paid $4.50 a day [!]

The *Vireless Vissel Yerker* would operate somewhat as follows: the hooker would carry with him a small portable transmitter capable of sending out keyed CW [carrier wave]. At the donkey would be located the receiver which by means of relays would blow a horn or actuate the steam whistle by solenoid. If we do get such a rig in operation we will sooner or later be tackled by half the loggers on the coast all wanting it. These blighters are quick to catch onto anything like that.

Kuchinka offered to pay part of the research cost and we set about madly shooting drawings back and forth, but we gradually became distracted by less interesting but more pressing matters and it was twenty years before "Talkie-Tooters" finally put the *vissel bunk* out of business in the BC woods.

THREE

The Waco

BUSINESS BOOMED. Until gasoline rationing came along and I had to tie up the boat. I would travel up the coast by any means I could—usually Union Steamship, but the Union Steamship, and the CPR, became just terribly crowded with troops going up to the big military installations at Port Hardy and Coal Harbour and Yorke Island—you couldn't get a stateroom, you were lucky if you got a chair to sit on. We had then taken on the responsibility of looking after a chain of little low-power radiophones the government had put in up the West Coast of Vancouver Island and also up the mainland, all the way to Prince Rupert, called ADC (Aircraft Detection Corps) They would sign up storekeepers' wives and so on, install this little battery set, and this was for reporting enemy aircraft. The silhouettes of the various types of Japanese bombers, I can remember them very well, they were always posted alongside these sets, so any time an aeroplane went over, of any kind, some panicky person would run in and report that the invasion had begun. This was considered very essential for the entire west coast of America, that these sets be kept working, and we were the ones who were supposed to go around and keep them working. At the same time we were selling six to ten of our radiophones a month and these had to be taken everywhere from Nootka to Bella Bella to be installed. So by the end of 1942 we were just desperate for a reliable way to get up the coast.

This was just the time when my cousin Rupert reappeared on the scene, working as test pilot for Canadian Pacific Overhaul in

Vancouver. The war had been good to him, and he'd been kept very busy working as an instructor on the prairies, then as a pilot with the Trans-Atlantic Ferry Command, flying warplanes over to Europe for service. He'd cut quite a swath and made a lot of money, but now he was back with a new wife, working at a more-or-less steady job.

I happened to see Rupert in action one day when I was driving back from a radio job at Steveston with Glenys and Ronnie, and wrote about it to Dad:

> Just before crossing the bridge to New Westminster we stopped at the big new CPAL aircraft factory and parked in a lot with provision for about 200 cars. The second berth in the reserve lot was labelled "R.B. Spilsbury" and Rupert's car was there so I went into the guardhouse and made enquiries and found that he had just taken a plane out on the river for testing and would be away two or three hours. We drove on to Westminster and out to the end of a dock near the BCER depot in time to see a big white Stranraer flying boat coming down the river from the direction of the bridge and take off into a westerly wind, circle the port at about 1000 feet and then head right off into the afternoon sky towards the gulf. Two hours later we were just getting out of the car at the house and we looked out and saw the same plane going over in the direction of Westminster again. Saw in the paper later that he had found a swamped sailboat in Howe Sound, alighted, pumped it out, and towed it to shore with the plane! He is very busy now with a lot of planes coming off the line and as he is the only pilot in western Canada licensed to fly their planes he can't get away very much.

He didn't have to keep very steady hours at CPAL and between jobs he'd wander down to our little shop and hang around, regaling us with wild flying stories and preaching about the great things aviation was going to do for the world as soon as the war was over. Like a lot of the oldtime barnstormers, he couldn't stop trying to sell aviation to everyone he met, and it didn't take him long before he was trying to sell it to us.

My cousin Rupert (white coat) with his flyboy pal and used-aeroplane dealer Wally Siple.

Rupert and the Waco.

Myself (left) and Jim Jacobson on our first and last charter for H.R. MacMillan, Kildonan cannery.

"Gee whiz, why don't you get an aeroplane to do your work up the coast? You can buy them a dime a dozen now because everybody's been grounded for the duration. But I think you might get permission — this is an essential service."

I told him we were having enough trouble getting gas for a plain old boat without attempting anything as exotic as an aeroplane.

"How much gas do you burn to go a hundred miles in your boat?" Rupert asked.

"Twenty or thirty gallons."

"There you go," he said. "A little four-seater on floats would take you the same distance with a thousand pounds of freight and only burn three or four gallons."

"I don't know anything about running an aeroplane, and I don't have time to learn," I said.

"Time?" he said. "Time is what aeroplanes are all about. An aeroplane would save you so much time you wouldn't know what to do with it. Anyway, I could run it for you."

It listened wonderful.

Ever since I'd flown up to Englewood on the CPAL flight with Tommy Lowrey I had been daydreaming about how much easier our work would be if we had an aeroplane at our disposal, but it hadn't got beyond that. Now, with Rupert available, the idea suddenly seemed to have a lot more credibility. The longer we went without the *Five B.R.* the more credible it became. You could get boat gas in dribs and drabs, and some service boats were putting up trolling poles so they could pass as fishboats and get the more generous fishermen's rations, but I didn't think I should have to do that. There was another factor at work here as well. It had become very clear by this time that our Vancouver end was where the heavy action was, and I didn't see how I could go on justifying leaving the office for months to float around doing nickel-and-dime repairs up the coast. The installation work was still up there but the deals were increasingly being made through the big logging and cannery bosses in Vancouver. A way of getting out and back on installations more quickly would allow me to stay in town and hold up my end as senior partner. Glenys and I bought a house in Vancouver in November of '42.

The next time Rupert brought the subject up, I said alright, if we did decide to get a plane, what would it cost? Where are all these bargains? He picked up a magazine, flipped to the used equipment section, and here was a guy named Albert Racicot in Montreal advertising a Waco biplane with wheels, skis and floats for twenty-five hundred dollars complete. I had never heard of the Waco make before and thought it had rather a tinny ring to it, but Rupert assured me it was quite popular and had a very good reputation as a small bush machine. The local Aero Club had kept a highly-prized Waco in service at the Vancouver airport for many years. Just the same, twenty-five hundred dollars wasn't quite an impulse buy in my book. This was a major investment. It had taken ten years to pay off the twenty-five hundred dollars I owed on the *Five B.R.* But Racicot hadn't had a live bite on the line for a long time, and he wasn't about to let us get away without a fight. I must have backed out three times, but he kept coming back until he'd not only agreed to take five hundred down and the balance over two years, he said he would fly the thing out to Vancouver for us.

That was the easy part. The whole reason his plane was for sale was that wartime restrictions had closed down the skies. How do we get around this? Rupert, in his breezy flyboy way, thought it should be easy but he could give me no actual direction on how to go about it. The only commercial planes that were flying were those of Air Canada, then called Trans Canada Airlines (TCA), and Canadian Pacific Air Lines (CPAL). TCA had no problems with restrictions, being a creature of federal government transportation policy itself. It had been created by act of parliament in 1937 and began flying with two ten-passenger Lockheed Electras and a de Havilland Dragon Rapide that Transport Minister C.D. Howe had bought, pilots, mechanics and all, from Canadian Airways in Vancouver. By 1942 TCA was flying over the Rockies with a whole fleet of fourteen-passenger Lockheed Lodestars.

CPAL was an even more recent creation than TCA, and also packed a lot of political clout, having been created when ten or so bush operations across the country were pulled together by the mighty Canadian Pacific Railway in 1940–41. CPAL were flying Vancouver–Prince George, also with Lodestars, but on the coast they were doing little more than just keeping their licences alive. Most of the time they had only one aircraft in service, two possibly, but one flying, and they were Dragon Rapides. They had licensed services to the west coast of Vancouver Island—mainly Zeballos and Port Alberni (Tahsis was only just coming into being), then up the east coast they had Alert Bay and one or two other ports of call. I don't think they averaged one flight a week. Possibly, in season, two flights a week. Their flights frequently wouldn't take off, wouldn't go if the weather wasn't right. Tommy Lowrey was their chief pilot at that time and he'd fly up somewhere or other like Alert Bay and hole up there for a couple of days, you never knew. He could only carry five passengers anyway. They later got the Barclay-Grow, which was larger, a very modern-looking plane for those days, but its undercarriage was magnesium. In the salt water it just fizzed. They only lasted a matter of weeks and they had to pull them out of service. No one else was operating at all.

In retrospect, the commercial flying which had gone on up until the war had been quite limited compared to what came later—there were more miles logged in five months after the war than in five years before. In the twenties the equipment was still quite limited—the old open-cockpit biplanes were used a bit for forestry and fishing patrols and aerial photography, but they were unsuited to hauling much in the way of freight or passengers, so the action was spotty—more of the barnstorming variety. By the time the thirties rolled around, some darn good equipment was available, but nobody had any money. Commercial flying was limited to the places that had no other alternative at all—like the northern bush, but while bush flying got a lot of headlines, it was still a pretty limited operation. The total of all airmiles flown by all the bushpilots in Canada before the war was laughable in postwar terms. On the BC coast, the hotspot of the pre-war years was the Zeballos gold rush of 1937–39. Canadian Airways, Bill Holland, Molly Small, Grant McConachie and Ginger Coote all piled in there for all they were worth, flying head to head, and they lost some lives doing it. But most of those smaller companies were swallowed up by the CPR, and for all its vast wealth even the CPR seemed unable to buck the government's flying restriction very successfully during the war. How could Spilsbury and Hepburn expect to succeed where Canada's largest corporation failed? Anyone who knew the score wouldn't have even tried. But I was lucky. I was young and I didn't have a clue what I was doing.

Well of course it was a real red-tape jungle, but we had one quite effective lever with which to pry open doors, and this was the ADC contract. We had to have the plane to service these little Aircraft Detection Stations, so we claimed, and this was a matter of national security. I had dealt with the wartime bureaucracy enough trying to pry loose restricted radio materials to know this was the approach to take. Instead of fighting them like most guys did, go right along with all their war talk but go them one better—claim that your military reasons for whatever you want to do are better than their military reasons for not allowing you. Anything that could be connected up to any of the so-called critical industries — fishing, logging, mining, farming — could be said to serve a military function if you looked at it the right way. Of all times, communication was needed badly then to support these industries — they were starting to put radio sets on all the fishboats, the tugs already had them, the camps couldn't get out the wood if they spent all their time waiting around for parts orders to get to town by slow boat — we had to get up and down the coast and keep these people going. It was difficult for anyone to contradict us because everything was hush-hush and nobody knew for sure what was going on.

I first of all went to the Department of Transport, and they said, "No way, not unless you get permission from the Air Force." So then I found out who to write to in the Air Force, in Ottawa, and given our very special circumstances they expressed willingness to let us fly in western skies, but they pointed out we'd need permission from the oil controller. I wrote to the oil controller and he said no, I would have to get permission from the DoT, Civil Aviation Division, and from the Air Force. It was a grand run-around. I closed the circle finally by getting letters from each and sending them around to the others. I bust a gut working at this. I just refused to be put off. In the end we got a special letter from the Dominion Government permitting us to fly with absolute freedom along the coast and to receive unlimited gasoline.

This was my special contribution. Rupert planted the idea and others came along to save our bacon later. But I wangled the permits that got us flying in wartime skies when nobody else could get off the ground, and that was what got the thing started.

While all this was going on our friend Racicot was getting terribly anxious, wondered why all the delay. The five hundred dollars was a bit of a problem — Hepburn didn't have any money and I had damn little, but I did have a life insurance policy I'd been paying on since I was very small and I was able to raise five hundred on that. Which gave us the down payment — we didn't worry about the rest, we'd make lots of money when we got the thing.

So now it's Racicot's move. The floats arrived by rail. That was very exciting, but where's the aeroplane? Weeks go by. Our customers all up and down the coast are jumping up and down, wanting to know when we're going to be delivering their equipment. No word from Racicot. The skis arrive by rail. We unpack those. We get all excited. Eventually comes a CPR telegram from Princeton. He had got that far, he was socked in, he could not get over the Coast Range. The Rockies had scared the hell out of him but the Coast Range was something else again. He'd never seen clouds so full of rocks in his life before. He had fourteen children and a wife in Montreal, he'd lost three weeks already and he had to get back. So he left the key with the stationmaster and got on the train. Rupert found the plane under a tree.

As soon as it arrived in Vancouver we had Cecil Coates of Coates Aircraft put the floats on and make up beaching gear so we could haul it up the ramp. We staked it down on the grass, we put a barbed wire fence around it — that was our only hangar for the first year. It was late in 1944 before we were allowed to share space in an old CPAL hangar with the Aero Club.

On December 20 Rupert had time off, so I took Glenys and Ronnie and met Rupert and Hepburn and three other friends at the airport to see the launching. It was a very handsome-looking plane, we thought, except for one thing: the colour. It had a very unusual paintjob, for a plane — black fuselage and bright yellow wings. It looked like a giant bumblebee. We got it all ready but the fog was too thick to take off. We waited around till noon and went up to the old White Spot on Granville for lunch. At 2:30 the fog started to lift and we went back and rolled it down the ramp, gassed up, and Rupert ran the engine up, but the fog started to get worse and by 3:30 we had to give it up. The only one who wasn't disappointed was Ronnie, who got to sit in the cockpit for a minute and was highly impressed. He dubbed the plane "keebird" and spent the next several weeks climbing the shrubs in our yard till they bent over and broke, saying he was up in an "airblane."

I think the first day we had clear weather we went on our first flight. On January 4, 1944, we left Vancouver with a complete radio-telephone station which we had built for a camp operated up Salmon Arm by Brown & Kirkland. A famous flight. Rupert was pilot and brought along his brother-in-law, Norman Hope. Hepburn went along as technician, and I went along to climb trees. So there were four of us; the radio equipment weighed approximately two hundred pounds; we had two large six-volt batteries weighing well over a hundred pounds; we had a Johnson Chore-Horse charging plant which weighed about 150 pounds, plus antenna wire and toolkits. Somehow or other Rupert got us off the river, got airborne, beautiful day, it was just a tremendous thrill.

It was a successful trip. Mr. Kirkland of Brown and Kirkland was there himself and he was terribly impressed and thanked us and shook hands, gave us a cheque for the whole thing, and we climbed into our little aeroplane, all the loggers went down there to push us off, we started it up, and away we went back to Vancouver, extremely pleased with ourselves. "The Fraser was so full of driftwood with the Xmas tides that we couldn't take off till 2 P.M." I wrote Dad. "However, we were finished the job and back in town by 5 P.M.. We were 25 minutes in the air each way. That trip would have taken two full days each way in the boat!"

Our euphoria came to an end rather abruptly when we landed and found waiting for us Mr. Carter Guest, regional inspector of air safety for the Department of Transport, and his assistant Mr. "Uppy" Upson. They didn't have much to do, what with commercial flying shut down, and the Air Force not being much interested in them, so our appearance on the scene was a big deal. They wished to know what was going on. This was the first they'd heard of us getting an aeroplane. We'd failed to notify them. We'd gone to Ottawa, over their heads, and their noses were very much out of joint.

They demanded our Certificate of Airworthiness, but that was alright, Rupert knew where to look for it in the glove compartment, and there it was, a dirty yellow piece of paper granted in Montreal—French-Canadian style. It was valid, I think, until September. The only problem was the weights it had on it, tare weight, fully loaded weight, net weight and so on. If you figured it out, it wasn't anything like I had been told. I thought we had an aeroplane that'd go 125 mph, pack four passengers plus a thousand pounds of freight. And we did. But not with floats. As it turned out from the C of A, if you added the float weight and filled the tanks with fuel, the speed fell to 85 mph and there was no weight left over even for a pilot. You had to leave out about a hundred pounds of fuel just to get the pilot in, never mind passengers and cargo. This was just the kind of thing Mr. Guest was looking for of course, and he raised particular hob with us. If it ever happened again they'd cancel everything including Rupert's pilot's licence. Furthermore, they advised us that when the C of A expired the next September they wanted a full inspection. So we had to be very cautious after that.

Not that we curtailed our use of the plane in any way. We were just cautious. On January 10 I took my next business trip, leaving

Sea Island at 10 A.M. to fix a station at Tipella up Harrison Lake. The trip was over a hundred miles each way. We took 1.05 hours up and .55 back. Glenys had waited lunch for me. A trip which had previously taken two days, completed between 10 A.M. and lunch. On April 12 we flew up to visit Mother and Dad at Savary, then down to St. Vincent Bay to work on a radio for Frank Kuchinka, who was still agitating at us to build the vireless vistlebunk. At the end of April I had to make a trip to the west coast "the old-fashioned way" by steamer, because Rupert was too busy to fly for us. This became a more frequent problem and I began to use Bill Peters, a TCA-CPAL man who was working his way down. He had problems, but at this point he was hanging on with the airlines part time. At the end of July we got off on another trip to the west coast with a couple of passengers bound for Ucluelet, but it was too foggy to land so we put them down on a seine boat in Barclay Sound. As we were tied alongside, a call came over the boat's radiophone for me. It was Hepburn calling from the *Five B.R.* up Jervis Inlet. He wanted to see me about a radio-telephone station there, so in an hour we were up Jervis, having a beer in the saloon of the *Five B.R.*. The next Thursday we were out again, as I wrote Dad:

> Our first call was Hardwicke Island to dump a passenger. Then back to Blind Channel. Then into Forward Harbour to Western Logging Co. to fix the station there. After lunch we proceeded to Acteon Sound. Fixed the station up there and had supper. After supper we proceeded to Mereworth Sound, Seymour Inlet and spent the night at Dumaresq Brothers' camp. It was lovely there. Jake [our new helper] caught a cutthroat right off the camp floats. I worked on the station till 4:30 that afternoon and then we pulled out. We descended at Village Island, Knight Inlet on our way back where I spent an hour making an adjustment to their set and then continued. We were about 5000 ft. when we went over Savary at 7:30 Friday so I guess you didn't even hear us.

For me, this was incredible, after all my years of chugging around the coast at 7½ miles an hour. Blind Channel should have been three days from Vancouver, Acteon Sound five days and Seymour Inlet a good week by the course I normally took. To visit all these places normally took at least a month out of my life. Placing them suddenly minutes apart was like travelling in time. It was uncanny. It was every bit as breathtaking as the breakthrough I had made years earlier, when I realized I could bring instantaneous communication to all these places through radio. I knew now that flight could be a huge boon to the people of the coast, and again, as in radio, I had the feeling I was uniquely positioned to provide it to them. Forget electronic depth sounders, forget wireless whistle punks—this had the potential to be much bigger than anything we'd come up with before.

The year 1944 started mild. On January 26 I was writing Mother and Dad to say spring had hit the garden at our new home on 2925 West 28th. The daphne was in bloom, the bushes were budding and the bulbs were two inches above the ground.

> I was so concerned over the progress the apple tree was making that I phoned up Keith [Dr. Keith Johnson] and he came over with his long pruners. Cut an amazing amount of wood but now it looks as if we hadn't done a thing. I had to move my pile of alder stove wood off the flowerbed into the basement as things were wanting to come up under it.
>
> Glenys got Mother's letter the other day. Also Ronnie's snow suit. I wish you could have seen him in it—tickled to death, and frightfully self-conscious about it all. Glenys is still trying to buy him underwear without success. She is talking of cutting some of my old shorts down for him. Poor fellow. Fortunately she is quite well stocked with infant's things for the new one, because these are all practically impossible to buy.

On February 6 we started out for Jervis Inlet in the aeroplane but were forced out of it by thick weather, so we decided to go over to Savary instead. We ran into more black clouds off Powell River and had to turn back. Rupert's wife Rene was along for the ride—the first time she'd been up with him, which I thought odd. She was a bright, vivacious woman who seemed to be keen for all kinds of adventure. Already she had talked us into hiring her brother Herb at the shop—she had an endless supply of brothers and sisters, many of whom would eventually become involved in my life to greater and lesser degrees. Just north of Buccaneer Bay we observed a large oil slick about two miles off shore. It appeared to be coming up from under water. A Kittyhawk fighter plane had been lost over the gulf the previous day and we were pretty sure this was it, so Rupert notified the Air Force when we returned. They sent a boat up to the site and evidently found what they were looking for, but of course it was all hush-hush and we never heard a thing about it. I suppose nobody knows how many planes they dropped in the course of their horsing around here during the war, but it seemed there was another search on every time you looked up.

In March the weather turned cold again with violent northwest winds, freezing the ponds and ditches and heaving the ground with black frost. The garden, which had been going great guns, was laid to ruin. In May our second child, David, was born. Like all second things, there was much less panic and less excitement than the first time around and Glenys resisted all my attempts to bring in help to see her through the first weeks. We hired one woman briefly, but Glenys sent her away after the first day. I was so darned busy I couldn't be of much help myself, but she was up and around in a very short time. She wrote Dad, "David is very good, wakes regularly for his bottle and only cries about 10 to 15 minutes out of the 24 hours—Ronald is still quite a handful but I am hoping he will get back to his old routine. He is very pleased with David and wants to help do things for him. David is very much like Ronald but has the same shaped head as Jim."

We were so busy at the shop it made previous years look like child's play. Much of this was directly due to the aeroplane, which raised our profile and greatly increased our effectiveness. We picked up five big installations on the west coast from the Dominion Government Telegraph, at Winter Harbour, Coal Harbour, Kyuquot, Port Albion and Nootka, worth two thousand dollars each. The Army's Navy (RCASC Water Transport) hired us to install thirty-five radio-telephones they'd bought from Marconi, after the CO noticed one of our posters with a picture of the "Keebird" on it and declared, "Those are the people we want!"

We kept putting on another man, then another, until one day I counted them up and discovered we had seven employees. Rene's brother Herb was turning out to be a prize find. He hadn't any particular background in radio, in fact he'd been a pharmacist. His only experience was that he'd built a home-made hi-fi set. But he was taking it up so fast I was able to send him out to the west coast in my place while I stayed home to be around Glenys. Within six months he was designing new equipment for us—a little semi-portable transmitter we marketed as the AD-10. It was a hit. We sold hundreds of them.

But we were still feeling short-staffed. We knew we couldn't find any more radio men so we advertised in the want ads for a flunky. The first response came from a scruffy-looking character who was middle-aged or more, had the face of a hard drinker and a certain displaced air that made you feel you shouldn't ask too much about his past, but we hired him anyway. His name was Charlie Banting.

We were even more short-staffed in the "aeroplane division" as the boys were calling it, never having a pilot when we needed one. Rupert was now quite busy with his testing, Bill Peters was back flying CPAL Lodestars into Prince George and we weren't able to find anyone else. In April I wrote Dad that I had a new man on the string in the person of Gordon Hollingsworth—Boeing's head test pilot—but the evening he was to start he phoned at the last minute to say he had been called back by Boeing. We put an application for a full-time pilot in with Selective Service, but were told our chances were slim. I found myself going up the west coast on the steamer again while the Keebird stayed tied down in her barbwire compound at the airport. I almost got to use it for the trip back, but Carter Guest and his sidekick came snooping around just as Rupert was readying the plane and saw him dipping the fuel tanks with a stick. This allowed them to deduce that the plane, like most of its vintage, had no inside fuel gauges. This was contrary to some new rule they had in their book, so the plane was grounded. It wasn't much of a job to fit the plane with gauges, but it was enough to put me back in the lineup for a steamer ticket home. The Waco was practically the only thing Guest's office had to think about, and we had to put up with a lot of attention from them.

Nevertheless, we did a lot of flying. In fact we did much more flying than we had planned, and this came about in a way quite unexpected to all of us. Our friend Mr. Kirkland of Brown and Kirkland-Universal Timber Co. called us one day and said he had to go up and cruise some timber, and could he charter our aeroplane? I talked to Rupert about this and oh, he was all for it and said, "You should get forty-five dollars an hour!" Kirkland was agreeable and the charter was very successful, and oh boy, it was some real money. We got two, three hundred dollars in, without even doing anything. In August we received a request from no less a personage than H.R. MacMillan, the lumber tycoon, accompanied by J.V. Buchanan, the head of Canada's biggest cannery concern, BC Packers. We got as far as Kildonan on the west coast and MacMillan wanted to go to Prince Rupert, but the pilot didn't like the look of the weather and turned back. Old MacMillan got a little huffy but we made him pay up at the full rate. If I realized at the time we were making an enemy of the most powerful man on the west coast I might have had some second thoughts, but at the time it just felt good to have a bit of MacMillan's loot in my hands. To make the same amount in radio took months of persuasion, days of shop time and more months of followup. We'd never *dreamed* of making it so easy.

So from then on, more and more charters. We just hadn't imagined the amount of aviation business that was lying dormant there and everybody screaming for it. Everywhere we went in the radio business they said oh, gee! If we only had aeroplanes! If we could only fly backwards and forwards to Vancouver. Get our men up here, supplies up, and all the rest of it. And airmail. They wanted everything. There was no commercial flying on the coast as we know it. Bill Sylvester, the founder of BC Airlines, bought his first red Waco just about the last days of the war, but until then we were alone in the field. With all our radio contacts in the logging and fishing industries, we were overwhelmed with demand for commercial flying right from the start. They wanted our radios, but when they saw the plane we brought the radios in, they wanted the plane even more. We very quickly found ourselves with the tail wagging the dog. Overnight the aeroplane was pulling in more revenue than the radio business. And as soon as we realized this, we started pushing it. We printed up stationery with a large image of the Waco spread across the page. We put out bright red posters with a photo of the plane pasted on—"Aircraft Charter Service! Waco Seaplane! 100 Miles An Hour! It Pays To Fly!"

I can only assume this conflicted with the terms of our permits, but you know, in making out reports, choice of words is everything. There was reference to the strategic value of the lumber, the urgent need to locate more aircraft spruce, this sort of thing, and we had no immediate complaints.

I never once recall ever having gone up and fixed an ADC station with the thing, however.

FOUR

One Damn Thing
After Another

MEANWHILE, CARTER GUEST HAD NOT FORGOTTEN US. True to his promise, there was an inspector to see us from the AID (Aircraft Inspection Department) the very day our Certificate of Airworthiness expired in September of 1944. Norm Terry, very good inspector, very well known, said where is this aeroplane? Terry, like Carter Guest, like Uppy Upson, was a flying veteran from World War I. So was Punch Dickins, the general manager of Canadian Pacific Air Lines, so was Romeo Vachon at the Ottawa end, and so many of the aviation fraternity of that time—they made a very tight little group that we stood no chance of breaking in on. We were outsiders. We would pay for our intrusion. Norm Terry walked up to the Waco, climbed through the barbed wire, took a very sharp, long-bladed fish knife out of his pocket, went under the wing, stabbed his long blade into it and made a big slash. Took out a notebook, made some notes and took another great slash. Made some more notes, went over to the next wing, did both lower wings, then he opened up both upper wings. And then he cut a great long slash in the fuselage, and he looked in there and noted that there was rust on the tubing. Went away, totalled it all up, came back and delivered the verdict: we had to replace three of the four wing spars; at least fifty percent of the ribs—these were of spruce—had to be replaced, all fabric had to be replaced—it was what they call ringworming, you could poke it with your finger and it cracks in circles because the dope has petrified—and the engine was due for an overhaul. Left us the list of things to be complied with, and when you've done this come back and we'll inspect it again.

Looking at our pretty little aircraft, that had become such a key to our financial security and future hopes, sitting there in tatters looking like an absolute worthless wreck was quite a comedown for us, but Terry had done his work with such casual assuredness we kind of shrugged and said to ourselves, well this must be a fairly normal sort of occurrence in this world of flying, now let's go over to Cecil Coates Aviation and see what it's going to cost us to get out. They gave us a minimum estimate, which they wouldn't guarantee, of ninety-two hundred dollars.

This was possibly the worst set back I can ever remember. We still hadn't paid Racicot his twenty-five hundred.

We then went around scratching our heads, trying to come out of the shock and wondering what to do, and I forget how, but I came in touch with Stan Sharpe, who was then operating Brisbane Aviation, just in the same area — which is now Vancouver's South Airport. Brisbane Aviation was a flying school, but during the war they were training mechanics for the Air Force. Stan was a sportsman, he was damn good to us, very sympathetic to our predicament and very anxious to take a job away from Coates. This is what transpired: he badly needed an aeroplane for his students to work on. It was agreed if I paid for materials he would supply the labour free, bringing our cost for the overhaul down to around forty-eight hundred, about half of Coates' estimate. A bargain, you see. There's nothing to egg you on like a bargain, even when you're getting into something way over your head. It seemed a godsend, but forty-eight hundred dollars was still more money than Hepburn or I had ever had in our lives, probably combined. But we were too deeply invested now, we had to get this money. So where?

Island Net to the rescue once again.

I'd met Jack Tindall over ham radio years before, and for the previous fourteen years he'd owned the store at Refuge Cove. Jack was far more interested in his hobby, in radio, than he was in running the general store by this time. He was terribly keen to get into the radio business.

I went and had a talk to Jack and he agreed he'd put the store up for sale, and if successful he'd come down and buy into the firm of Spilsbury and Hepburn. The price was five thousand dollars. Rupert's wife Rene came up with two more brothers to buy the store — Norman and Doug Hope. Jack Tindall moved to West Vancouver, and in November of 1944 we went to Stan Sharpe with a cheque for forty-eight hundred dollars. We were ready to go ahead, again. In January Bill Peters had been laid off again, so I quick tied him up by offering him a job working full time for us.

All we had to do was wait for Sharpe to get the plane repaired. I went out to the airport every week and it was so strewn around the shop I couldn't even find it. Completely stripped apart. They were doing a terribly thorough job. All the tubing was being sand blasted and it had to be treated, and all the ribs were being formed up on jigs on great big long tables with dozens of students working on it, and all glued together and varnished and it went on and on and on. The year ended, February 1945 rolled by, then March. It was killing us to keep an unemployed pilot standing around so I somewhat apologetically approached Bill Peters about filling in as a radio salesman. He shrugged and went along with it. But we needed the aeroplane. Our work upcoast was at a standstill, and we were badly in need of money. More weeks went by and it was very difficult to see any progress. It got to be April. Stan Sharpe, for all that his heart was in the right place, just wasn't getting the job done fast enough. His aeronautical engineer was a very clever guy, but not practical. I could see that. I was worried sick, dashing back and forth to the airport, putting off customers—"yes yes yes, we'll be up there with our aircraft, we'll do this, we'll do that"—and no money was coming in, payrolls rolling around, and things were looking really bad.

Only a miracle could have saved us at that point. I mentioned before that we had this flunky, Charlie Banting, who had come to us through an ad in the papers. Charlie reminded me very much of Will Rogers in his appearance and speech, and he had the same sense of humour. Very dry. I grew to just love it. Well, Charlie was useful. He would never know much about radio, but he drilled holes in ribs, strung wire, swept up the shop and did all we asked of him.

One day in late April as I was just about to set out on another of my despairing forays to the airport, Charlie said, "Do you mind if I come out with you, I'd like to see this aeroplane."

I said, "Sure Charlie, are you interested in aeroplanes?"

"Well, a little. I'd just like to have a look at it," he said.

I asked Hepburn, he didn't need Charlie that afternoon, so we jumped in my old 1930 Plymouth and we rattled all the way out there. Stan Sharpe came out to meet us and we went all around the Brisbane hangar together. Charlie didn't say a word, just walked around looking at everything—I didn't even introduce him. Stan

The Waco at Nootka cannery.

Charlie Banting.

Bob Gayer. Would you trust this man with your new aeroplane?

Bill Peters with a planeload at Goose Bay cannery, Rivers Inlet.

showed how they were coming on with the wings, how the crew were working on the fuselage, somebody else had the engine pulled apart—I couldn't see much progress, and Stan just said well, I think it'll be coming along pretty soon now... Then he pulled me off to the side and said, "You know I'm getting pretty worried about this job, I know I made a deal with you and I want to stick to it, but this is coming close to breaking me!"

This shook me up pretty good—the whole thing was starting to get a black, hopeless feeling about it and I'm afraid I wasn't in very good humour driving back to Cardero Street. The aeroplane had now been grounded for seven solid months. Charlie didn't say anything for a few miles, then he turned and looked at me over his glasses, with his big sad eyes, and said in his slow way, "You know, them fellas ain't never gonna get that aeroplane flyin."

"What makes you say that Charlie?" I asked, a little bit irritated, but curious that he would venture such a comment.

"Hell, they don't know what they're doin," he says. "Two years from now they're still gonna be workin on it."

"Is that so? What do you know about it?" I snapped.

"Well, dammit all, I wasn't going to tell you that," he said. "I shouldn't tell you. But I guess we're into something here bigger than both of us." And so saying, he reached into his inner pocket and dragged out his wallet and handed me one of the dirtiest, most worn little pieces of paper I'd ever seen. It was an air engineer's licence, classes A, B, C, D and X combined. The A and the C were for routine aircraft maintenance and the B and the D were for serious overhaul on certain makes. The X was for overhaul on *any* make. He was one of Canada's most senior aviation mechanics. He'd been superintendent of maintenance for Arrow Airways, a large bush-flying outfit operating out of The Pas in northern Manitoba, later absorbed into CPAL. But drinking was his problem. And he associated his drinking with planes. So he quit his job, took his wife and daughter, moved out to Vancouver and decided he'd do some other kind of work, never again to go near an aeroplane. And by pure fluke ended up with us.

I turned around and drove back out again and introduced Charlie to Stan Sharpe, formally this time. Well Stan was just about floored when he saw that licence. There wasn't a handful like it in all of Canada. I said okay, this is the deal. Charlie will work here for you and I will pay his wages. It'll cost you nothing, but let him boss the job.

By June, a little over a month later, the aeroplane was out there with nine coats of hand-rubbed dope, just gleaming. Beautiful job. When it came to the weighing-in, somehow Charlie managed to get the tare weight down so the plane came out with a decent payload. He never would tell me how he did it. "Waaal, I had to simplificate a little and add some lightness. You ain't smart enough to understand," he would say. He was slick. The inspector just shook his head and signed it out.

Charlie had saved us, and his reward, much to the dismay of his wife, was to get more deeply involved in flying than he'd ever been, but that is getting just a bit ahead of ourselves.

We were understandably proud of our shiny new aeroplane, and wasted no time pulling Bill Peters off the sales beat and catching up on our much-delayed service calls. It was only then that we realized how successful Bill had been in his fill-in job. In one two-week period he had sold over twenty thousand dollars worth of radios, more than the whole company earned in some years! He was phenomenal. I should have left him selling and taken flying lessons myself, but he wouldn't have gone along with it. To a pilot, flying planes is the only job that really rates, and Bill was anxious to quit fooling around and get back to some real work.

On June 1 we sent our customers a circular letter on our new Spilsbury and Hepburn letterhead with the photo of the Waco engraved in gold and "Land...Sea...Air" written below it. "We are pleased to announce that our seaplane CF-AWK is back in service after a undergoing a major overhaul. This work has been done at Vancouver and embodies all the latest techniques in aircraft construction methods. In addition we have added new two-way radio and range equipment and also directional gyro and all blind flying instruments, which will bring the aeroplane up-to-date in all details... The use of the plane last year in speeding up our service to outlying points proved very valuable. We feel sure that it will be even more valuable in future as both our customers and ourselves take more and more advantage of this service."

We were a lot hungrier for charters than we had been, now that the damn plane owed us so much. As soon as Charlie had turned it out gleaming and sparkling from its overhaul, one of the first calls that came in was from this Robert B. Gayer, E.M. (Mining Engineer), who the previous year had inveigled me into the wilds of darkest Vancouver Island on a packhorse to install a radiophone at his mine. The call was personal. He wanted to talk directly to the president and nobody else. He was inclined to be mysterious about his purpose, and I guess I should have been more suspicious, but it had to do with renting our aeroplane and I couldn't afford to turn my back on that, and he knew it. He wanted it for one day — not over four hours flying time. That would be four times $45, or $180, which didn't deter him, but there was one condition. He only wanted the aeroplane — no pilot — and that was crazy, but he said he had his own pilot, and only this pilot would know how to fly the aeroplane for this very special job. I declined on the basis of contravening our insurance policies, saying that only certain pilots were approved. He said I had nothing to worry about since his pilot was about the best known and most qualified pilot at Vancouver Airport, and we had already used him to fly our Waco. He referred to Len Milne. Len was the manager and chief flying instructor of the BC Aero Club, and without question was fully qualified and known to be a very careful pilot. I checked with cousin Rupert and he sanctioned it. Rupert was in the midst of a heavy program at CPAL Overhaul and couldn't handle the job himself. Hepburn, once he saw the part about $180, cash, approved instantly, mentally calculating how much we would save by not having to pay our own pilot. So it was arranged that the aircraft would be signed out and ready at first light — just leave the key under the mat!

We didn't see the aeroplane again for about four days. In the mail, on the fourth day, I received a cheque for $180 from "Robert B. Gayer, General Manager, CANgold Mining & Exploration Co. Ltd." Hardly had we opened the mail when I received a sizzling phone call from Charlie Banting at the airport. He said, "Well, you got your aeroplane back, but it looks like it's been through the Battle of Britain. You better come out."

I went out.

Charlie had understated it. The outside of the fuselage and the floats were a uniform dull grey colour. Someone had rubbed around the registration letters so you could just distinguish CF-AWK. The inside of the cabin was an indescribable mess. The ivory-coloured head-lining was all mottled and ranged in tone from grey to black. Everything you touched left a greasy black smear on your fingers.

I was speechless. When I recovered, I got mad. Then I became enraged.

I went in to the nearest phone and called Gayer, at the office, at his home and everywhere else he might be, but the best information I could get was that he had had to leave town suddenly and no one knew when to expect him back. Even his wife Louise didn't know where he was, and apparently had no knowledge whatever of the aeroplane episode. I guess Gayer's pride didn't permit admitting to the schmozzle. It took Charlie with a helper a whole week to clean the aeroplane up, including installing all new head-lining and drycleaning the seats and cushions.

It was years before Gayer (I call him Bob now) told me the whole story. When Bob was operating the CANgold Mine, he had to take all his supplies on a small boat up Great Central Lake to the head where he transferred to pack horses. The lake freezes over most winters. The ice normally was not heavy enough to operate vehicles on, or even to cross with snow shoes. On the other hand, it was too heavy for his little tug to break through. It extended about four miles down the lake from the head, and if he could just find some way of breaking a path through it he could get his summer operation started several weeks ahead of time. He had heard what they did on Great Slave Lake, and talked to some of the people who were working up there. They spread lamp black on the surface of the snow. The sun's heat was absorbed, and they opened up a long canal in the ice way ahead of break-up. They spread the carbon from vehicles, and they said it didn't take very much.

Bob could not use vehicles on the ice on Great Central Lake, but he had a brainwave — why not spread it with an aeroplane, like crop spraying? Apparently the first obstacle, after finding someone dumb enough to loan him a plane, was to get enough lamp black. He originally thought in terms of many fifty-pound bags of the stuff, but when he came to buy it he could only get it in one-pound packages. He said he cleaned out every hardware store and paint shop in Vancouver, and then only amassed about fifteen pounds in total. He then borrowed two pillowslips from the trusting and long-suffering Louise, and decanted all the small bags into these. He cut a small slot in the corner of each pillow slip with the intention of squeezing it out smoothly to spread it evenly. He would ask the pilot to fly slow and low (about five feet above the ice) while he, Bob, would wedge the door open with his foot, squeeze the bag and paint the snowy desert.

That isn't exactly as it happened. First off, he had to get a young, inexperienced pilot. Len Milne couldn't make it. I suspect he bowed out of that one on purpose. Then, it seems that just when Bob wedged the door open (having determined by previous experiment that this creates a strong vacuum and sucks all the air out of the aeroplane like a vacuum cleaner) the young pilot got worried about not being able to see where he was going. The nose of the aircraft, with its big radial engine, was naturally coming up at such slow speed and obstructing his vision. So what does he do? He opens his side window and sticks his head out, of course. But this completely reverses the cabin pressure. A blast of air rips in through the door, exploding the bag of soot and distributing it evenly and instantly over every square inch of cabin interior, including the inside of the windscreen and all the instruments on the panel. It is all perfectly black, like the inside of a cave. Somehow, with his head sticking out the window, the pilot manages to make a landing.

Bob says the two of them looked like Amos and Andy. Only the whites of their eyes showed. They taxied back to the far end of the lake and got the gang to go to work with hoses to try and clean up, but the best they could do was to rub enough off the outside so the registration letters could be recognized. Bob never saw the pilot again. The young fellow was afraid Carter Guest would find there was a law against throwing soot out of an aeroplane and he'd lose his licence, so he vanished quietly from the local picture.

As for Bob, was he discouraged? Hell no, he signed up at the BC Aero Club, learned to fly, got his licence, bought his first aeroplane and started his own flying company, Associated Air Taxi, eventually forsaking mining entirely to become one of the important factors in postwar aviation in western Canada. In the next few years he did just about everything you can do with an aeroplane—except throw soot out of it.

One thing the plane did was extend our range. On the boat it just hadn't been practical to go further afield than the north end of Vancouver Island or Seymour Inlet on the mainland side, and our business had tended to remain confined within that area. But with the plane, the whole coast was within a few hours' run, and we began spreading out. On June 11 I made my first ever sales-service trip to Prince Rupert in the Waco with Bill Peters flying and Charlie along to keep things from falling apart. I wrote Glenys on letterhead from the Hotel Prince Rupert—"Headquarters for Commercial Travellers—Mining Men—Tourists."

Bill and Charlie are out doing the beer parlours. I
have just been down to see the train pull out. Got a

big kick out of it. Saw off the manager of one of the canneries on the Skeena. Takes him an hour by train to get there. Seems very funny. Thinking of pulling out for the Nass River in the morning to do Arrandale and the Wales Island station — weather permitting. Next day I think we will be doing the Skeena and calling at Carlisle and Claxton canneries. Then we will be returning to Butedale for a day or two and then working our way down into my old territory with quite a lot of calls to make. We arrived at PR about 4 P.M. yesterday (Monday) in pouring rain, heavy cloud and fresh SE wind. Landed at the RCAF base and tied up to a buoy. Two crash boats attended us and ferried us ashore. Then we made arrangements to pull the plane out on the ramp and got the tractor down etc. we all got into the water up to our waists putting on some beaching gear they had on hand, only to find after hours of struggle that their gear, which was built for a Norseman, was too large for our plane. We had to get towed back to the buoy again. It was pouring so hard that wading in the water didn't make us any wetter. Bill lost his overcoat in the water and then his $13.50 town shoes came apart and was he mad! Charlie and I laughed until we fell in.

Tonight the weather is lovely and I have spent the last two hours walking around examining Prince Rupert close up. Very interesting, and quite a pretty place. Whole thing looks very green. But their gardens are rather backward. One man was telling me that his cherry tree is going to have cherries for the first time in 12 years. Usually it just blossoms. They have just had six weeks with practically no rain and it is something of a record I believe. Do they ever boast about it too. But it was not much help to us as we missed it.

The little plane has been running beautifully and the trip going very well. Getting a lot of work done and making a lot of contacts. Charlie is very good. Acts as a steadying influence on Bill and is a great help to me around any radio job. Makes a good team.

We went as far as Wales Island and actually into Alaska, just for the say of the thing. The plane itself was about the best sales gimmick I'd ever had, especially up there—people were just so amazed to have anyone come all the way up from Vancouver to see them, they couldn't get over it. Once they saw the plane, Marconi didn't stand a chance of getting their business. We were away twelve days, and almost as soon as we got the plane unloaded Hep took it out to the west coast for a week. Then when he got back, I was waiting to take it up to Ocean Falls for two days. It was absolutely hectic all day at the office. I reached the point where I simultaneously wrote invoices, talked on the telephone, conversed with customers and dictated letters. It didn't bother me any more to have about six people waiting impatiently in the lobby. Of course this was special, as the fishing fleet was making ready to go out after the big run of pilchards that used to appear off the west coast of the Island in those days, and it seemed the fishermen all came to us whether we wanted them or not. I wrote Dad:

> By the way things look every week is busier than
> the last, so I don't know where it is taking us. There
> is no longer time to stop and consider where we are
> going or how we are doing. Hep was away last week
> so that threw the whole load on me. Now the jobs are
> piled so high and coming so fast it is confusing.

We took a ninth man on at the shop, which we had doubled in size during the spring to hold all the new staff. Charlie stayed in town for a few days to put in some time straightening things out around the work area, and what a difference he made. He mounted our small lathe on the wall, cut away the bench to make it more convenient to work around the drill press and got the place spotless. It was quite a boost to see how competent he was, and how dedicated he was to making things better, just on his own initiative. Between him and Jack Tindall, who was also picking up a tremendous load, it made things just about possible for me.

We found a few aeroplane charters—just enough to whet our appetites. Bill flew a diamond drilling outfit into Nahwitti Lake, about twenty miles inland of Cape Scott. We had another similar deal in the Chilcotin country—gold mines were becoming active again. Pacific Mills had us out to fly seventeen hundred miles of timber reconnaissance around Ocean Falls. One morning I was called out of bed at 7:30 by my old friend the Reverend Alan Greene, superintendent of the church hospital mission headquartered in Pender Harbour. He had a stretcher case that desperately wanted bringing in to the Vancouver General, so I got Bill Peters out of bed and dispatched him to Pender Harbour on a mercy flight. The round trip took an hour and fifteen minutes, including loading and unloading the stretcher. It was our first mercy flight — but not our last.

By July we had the pilchard boats pretty well cleared away so I took the plane north again to Acteon Sound, Simoom Sound, Nimpkish Lake, Rivers Inlet, and Talleho, just north of Ocean Falls. We took two passengers, a Mr. T.B. Jackson, who was manager of the big Pacific Mills pulp operation at Ocean Falls, and an assistant of his. Bill Peters took the opportunity to harangue Jackson about helping us start a regular service up the coast to the Queen Charlottes, and by the end of the trip had the guy practically promising to loan us fifty thousand to buy equipment.

Even though I was home most nights, Hep and I often worked until ten or eleven P.M. at the shop and Glenys was once again beginning to complain of feeling like a grass widow. The boys were booming along so that every time I turned around I seemed to have missed a chapter. David was now a year old and weighed thirty-one pounds, as much as Ronnie had at *two* years. It had been so long since I'd taken him up to see the grandparents I had to send them a written report on his progress: "He is walking quite well now but finds it hard work with his bulk. He is tanned like a coffee bean all over his arms and legs with white in the wrinkles. His face is very round and ruddy. He does a lot of talking, but no English words so far. He eats enormously and with gusto."

I announced that we had "actually decided to try and take holidays this year."

> The plans are to take the boat out to some sheltered cove where the kids can get out on the beach and try to take things really easy for a spell. Not that I need it, but it will do Glenys good. She hasn't been out much since we moved to town. When I get back from this next trip I will have a better picture of what to expect and will let you know. We have to consider the others' holidays. Hep is taking

the last 10 days in July. Mrs. Yates leaves tomorrow for her two weeks and Herb and Shorty go in August. Jack is too new to get any and so is the other fellow. Ditto Peters and Banting.

As usual, by the time I had everything else out of the way, summer was over, and we had to pass on our holidays for another year. After that I was afraid to even mention holidays around home. But it was true that I didn't miss them. I was so fired up with the work we were doing, especially in the flying end, that I really didn't want to be anywhere else.

Back on May 7, in the middle of our Waco crisis, there transpired a rather significant event which we hardly had time to notice, although it was to change everything. I wrote Mother:

> We woke Monday with the sirens wailing and knew what it must mean. Shortly afterwards the kids were running around shouting EXTRA. We worked all that day but it was difficult to get anything done as so many places closed up as the day progressed. They were not supposed to stop work but it was such a good excuse to rush out and get drunk there was just no holding it back. Downtown was bedlam with car horns and shouting and torn newspapers and toilet paper strewn all over. It was hardly safe to drive. They only killed one person though! It looked like they'd kill a hundred. Then Tuesday was declared the official holiday and VE-Day celebration and nobody celebrated, they were all played out. The city was quiet as a tomb.
>
> We listened to the King's speech at noon. We were working in the garden and I turned the speaker out the den window so we could listen to it as we weeded. The next door neighbours listened to it with us. The king seemed much improved in his speaking since the last time we heard him before the war.

What we didn't realize at this point is that we were being watched every move by the Big Brother of the west coast air business, Canadian Pacific Air Lines. We hadn't had the plane back in service long enough to get caught up on our own calls when we had a visit from a department of government I'd never heard of before, the Air Transport Board. It had been formed by C.D. Howe during the latter part of the war to regulate commercial flying, and we were now informed that if we were going to continue chartering we would have to obtain a Class 4 charter licence. The war was over and so was our special permit for the flying we did. In applying for this new licence, the catch was that we were by no means guaranteed of getting it, because anyone else could intercede and try to get it for themselves. We had no choice by this time but to plunge onward and apply, we were so far in debt, so Esmond Lando, our lawyer, got out a very formal looking document, shipped it away to Ottawa, and we began a nervous thirty-day wait to see if anyone would intercede. We made it to the twenty-ninth day and almost started celebrating, then a telegram arrived to notify us Canadian Pacific Air Lines was interceding. This was a blow to us naturally, but what annoyed us particularly was that we had not received a copy of their grounds for objection as provided in the procedure. Just this telegram mentioning something about a delay. I suppose it was deliberate on their part, to give us no time to respond or amend our application.

Finally their package arrived at our little office down on Cardero Street. I remember it so well. Very formidable-looking envelope, legal size, about an inch thick. Registered, insured, everything else, had to sign for it. Hepburn and I opened it up and read the covering letter, the table of contents, found all the different parts to it — but there was something wrong. There was all sorts of material that didn't seem to belong — original letters between CPAL's western manager "Transcanada Tommy" Thompson and Punch Dickins, the general manager in Montreal, discussing our business, discussing their strategy against us. Slowly it dawned on us that there had been an accident, and what had happened was, somehow in the rush to beat the deadline, some poor secretary had inadvertently slipped the entire CPAL file on us into this letter.

We had to think about this. We were very aware of the stakes. If we couldn't get this licence, we'd lose our aeroplane, we'd lose all the money we had tied up in it, we couldn't keep in business, we'd lose our houses, our cars, it would spell the *end*. The CPAL, it was clear, had been watching our every move with the Waco, carefully documenting each illegal charter, every transgression of ATB rules. They had amassed enough evidence to have us banished from the skies forever. But the whole CPAL case against us was here in this file, we were holding it in our hands, and unless we shipped it back to them, the slate was wiped clean.

We decided to be noble about it. If we sent it back, that poor secretary would be found out and lose her job, and we didn't want that. So we dropped the entire file into the furnace.

The hearing was held in the old courthouse, now the Vancouver Art Gallery, on September 2, 1945. The Air Transport Board came out in force and listened as Mr. Lando presented our case, then called the opposition. The CPAL was represented by a sharp young lawyer from Montreal, but without the essential background material contained in the lost file, he was completely at sea. In his embarrassment he wound up uttering statements such as one I recall to the effect that Port Hardy was an important coal mining center. There hadn't been a piece of coal dug at Port Hardy in a hundred years. In their last-minute scramble to replace their information, they'd found this in some old book. When the poor fellow was finished, the chairman of the board rather tersely adjourned and left. Several months later we were granted licence ATB 281 authorizing us to fly non-scheduled charter service from a fixed point in Vancouver "to any point or area that can be safely served, but more particularly along the coast line of the Province of British Columbia."

So here we were with an overhauled aeroplane, a charter licence, in debt up to our ears, but raring to go. Bill Peters didn't quite wear the wings off the plane but by the end of the year we did have the engine worn out. Both sides of the business expanded to the point our staff count now went up to thirteen. Demand for the plane was such that we began to feel very strongly the need for another aircraft, perhaps a larger, twin-engine machine that could be flown as an IFR (Instrument Flight Regulations) type of aircraft on a scheduled run. The Powell River Company and Kelly Spruce had come to us urging that we start a scheduled service from Vancouver to the Queen Charlottes and it was clearly the way to go.

Three twin-engine Grumman Goose aircraft came up for sale by the Air Force and I was very excited about the idea of getting them for the Charlottes run. The price was twenty thousand and some, far beyond our means, but I was becoming more audacious by the day. I got the idea I might talk my old friend R.M. Andrews into backing me. I'd met R.M senior before the war when he bought Twin Islands, just off the south end of Cortes. He was supposed to be worth $160 million. He'd made it mining around the world and had a large mining equipment concern, Andrews and George, with its head offices in Denmark and Tokyo, but he could see the war coming so he had a lavish estate built on this remote island and moved back to BC to wait out the war. Twin Islands became one of my regular stops, as the old boy kept me coming back to install radios in every one of his fourteen bedrooms, then a radio-telephone, then phones between the rooms, then a radio-telephone in his big yacht, etcetera. People knew he was close to the Japanese, so naturally the coast was soon alive with rumours that Twin Islands was a Japanese spy centre with underwater submarine docks and all the rest. The army came up and ripped out his big radiophone at one point, and I had to appear before the brass in Vancouver to explain why it couldn't be used to communicate secrets to Japan before they would let him put it back. But the traitor thing rubbed off on him, and people considered it fair game to take him for anything they could get. He often said I was the only Canadian he'd dealt with who hadn't screwed him. I thought this ought to be worth something, so I hit him up about going in with us on these planes.

I'm quite sure he would have done it, except for R.M. Andrews *Junior*. R.M. Junior had a private pilot's licence and the old man was trying to involve him in things and make him feel good, so he brought him in on our talks. In this case, R.M. Junior took his advisory role seriously and decided it was his solemn duty to save his old man from making a big mistake. He talked to a few experts and came back warning old Andrews to have nothing to do with these Grummans. He explained they were built in the thirties and were therefore obsolete, they could never compete with the newer equipment. R.M. Senior really didn't care about the details but he was very intrigued with my peculiar talents and would have liked to embark on a joint venture with me, but R.M. Junior dug his heels in, and rather than humiliate his son, the old man begged off. I looked around for other backers, then the planes went to the Powell River Company as executive aircraft. The same model, the Grumman Goose, went on to become one of the most successful commercial aircraft on the coast, forming the backbone of several airlines in BC and Alaska through the forties, fifties and sixties. The three Grummans R.M. Andrews Junior said were obsolete in 1945 were still in service in 1988. If we'd got started with those three excellent machines and a backer of R.M. Andrews' stature, the story of our subsequent flying operations, and Canadian aviation history, might have been quite different, but a bit of family politics got in the way.

We had to re-power the Waco, so we found a war-surplus Cessna T-50 Crane with two Jacobs L4MB engines and two Hamilton Standard propellers, which we badly needed. Hamilton, incidentally, was the man who built the Malibu Resort in Princess Louisa Inlet. He owned the patent on the variable pitch propeller, which he was supposed to have stolen from one of his employees. He sold propellers very successfully to both sides all through the war. Anyway, this aircraft was in Calgary and it had only four hours and thirty-five minutes on it since new. The price was two thousand dollars. I took Bill off the Waco and sent him and Charlie back to bring it out. I told them to be quick about it because the Waco was supposed to head north the following Sunday to start a big reconnaissance job up Kimsquit Arm, then the next week the Powell River Company wanted it for a timber survey up the Blue Water River a hundred miles east of Kamloops. After that, there were several days work for a mining company up north of Bridge River. It was awkward to take any time off, but I had given up waiting for a break in the schedule, so out they went. They estimated they would be back in eighteen hours. Between the red tape involved in getting the plane officially released by the Air Force and the bad weather over the Rockies, I didn't see them for two weeks, during which time the Waco stayed on the ground.

But this Cessna Crane was just a beautiful little machine. It was so beautiful, in fact, we said no way can we rip it apart just for the engines. Why not use *it* to start this scheduled service to the Queen Charlotte Islands, we thought. They had others not quite so new, so we bought one of them for $630 to re-power the Waco and went into the scheduled flying business with the good one. I wrote Dad:

> The new plane is a twin-engine land plane, 5 passenger, and quite fast — 140 mph cruise and 170 mph top. I'm afraid you won't see much of this one around Savary till you put in a landing strip though. We will use it chiefly for long range reconnaissance work and to start a passenger service to the Queen Charlottes using the new terminus built there by the RCAF. It will make that trip in 3½ hours. We will take on a new pilot for this work alone, keeping Bill on the Waco as much as possible. Then we will go about procuring another Cessna like the first, to act as a standby ship. We may fly as many as four trips a

week to the islands and the first plane will earn
enough money to buy a second one in about a month
of this work.

Our one act of preparation for this new stage in our aviation
career was to place an ad in the paper. I still have the original
cutting. The plan was nothing if not ambitious. "A twin-engine
Cessna took off today on a 'familiarization flight' launching British
Columbia's newest regular air service," wrote the Vancouver
Province in its October 4, 1945 edition. "Today's flight, in charge of
the well-known bush pilot Bill Peters, will land at Masset, Terrace,
Smithers, Prince George, Quesnel, and back to Vancouver... At
present Spilsbury and Hepburn has two Cessnas, but in about a
month two *Canso flying boats* will be added to the fleet.(!)" We got
three passengers, Peters flew the plane and I went along to shake
hands and tack up schedules. We refuelled at Port Hardy, dumped
our passengers at Sandspit—the Air Force turned out en masse at
Sandspit, they didn't know what the heck was happening—the
weather socked in, we never could get to Prince George, so we flew
back. Only to be met by an irate Carter Guest.

"What the heck is going on? You can't fly per seat mile on a
charter licence, number one. You have to have a scheduled flight
licence. You have to *apply* for one. Furthermore, the aeroplane
doesn't have a Certificate of Airworthiness." And he was right; it
didn't. We never even looked into this. "This is a training aircraft,
not approved, never has been approved for passenger flying," he
said. Another version was approved, what they called the heavy
wing version, but our model was purely for training. Instead of
spending the next month paying for another plane, it spent the next
year in a storage hangar. That was bad enough, but Carter Guest
wasn't finished; he also lifted Bill Peters' pilot's licence.

So we were learning. We did get a licence to use the Crane for
aerial photography, we chartered it out for that work later, but it
wasn't the answer to our scheduled flight hopes. Meanwhile we had
no pilot and all our work was tied in a knot again. Phooey!

FIVE

The Whistling Shithouse

MY FIRST JOB WAS TO FIND A PILOT to fill in while Bill was suspended and tackle this huge backlog of flying which had been waiting before the Cessna Crane episode. Rupert was busy ferrying war surplus flying boats down to South America with some Air Force buddies of his so we couldn't get him, but there were all kinds of ex-RCAF fliers dropping around wanting work. Most of the war vets weren't ready for commercial flying. I remember one very impressive-sounding chap, a wing commander. He'd been a pilot twenty-eight years, in both wars. He'd been commanding officer of one of the main seaplane installations for the last three years of the war and seemed to have all the qualifications we could ask for. We had flying work backed up in all directions, but in four days he didn't complete one trip. He had the skills, but he lacked what it took to go out on his own and move the loads.

One ex-service man stood out. He was well-spoken, had pre-war bush flying back east and seemed to have a good grasp of the business side of things. I hired him on a temporary basis and had him fly me out to the west coast to fix a couple of boats in Barkley Sound the same day. It was sunny and pleasant in Vancouver but we hit snow going over the Alberni Pass. Out on the Sound it was blowing a gale and by the time we got back up Alberni Canal the pass was closed right in. As usual, none of the hotels in Port Alberni had room, but we managed to get space in a flophouse through the chief of police, who was a radio customer. The next morning we awoke to find the hotel we'd been turned away by burned to the ground.

In the morning the pass was still blocked so this new pilot decided to earn his pay and spiralled us up through a hole in the cloud at eight-five hundred feet. This was strictly against the rules of so-called "contact flying," where you were supposed to keep visual contact with the ground at all times, but he knew what he was doing. As soon as we got on top we found clear sky over the gulf and we went right over. After we got back to the office we got a call from Victoria wanting to know if we had a plane lost. The BC Police up island had reported a plane circling in the mountains behind Port Alberni. I wrote Dad about the trip December 7:

> I had the new pilot with me. He is a very nice chap and I think we will be keeping him. His name is John Hatch. He is a quiet one for a change and has a very nice manner. He has had five years experience overseas in a Mosquito bomber and fighter squadron and had considerable prewar-war experience in Eastern Canada. Bill Peters has not received his licence back yet. As a matter of fact Ottawa has not got around to acknowledging it yet and it is nearly two months since it was sent to them!

John Hatch, or Johnny, as everyone was soon calling him, was a godsend. Not only was he a terrific pilot, he immediately sensed the potential of the flying business and set about doing everything in his power to develop it. He was tremendous on public relations. In no time at all he had Poldi Bentley, the head of Canadian Forest Products, wrapped around his finger. Maude Johnson up in Seymour Inlet was practically in love with him. He was the favourite among all the little camps and did us no end of good. He had a wonderful way with planes and with people—especially women.

The plane had never flown so regularly, or so smoothly.

Johnny even got the Cessna out for a few illegal passenger flights. I wrote Dad January 24, 1946:

> The Cessna went to the Queen Charlotte Islands and returned yesterday, making the trip down in 2 hours 50 minutes, a record for us so far. On his way up he had to fly at 12,000 feet to dodge bad weather and his temp. up there was 28 degrees below zero Fahrenheit! The Waco has been out almost every

day. You probably see it go over frequently, bulging with loggers and radio experts. Bill Eaton has just returned from Nimpkish today. Herb is in Zeballos installing a new station and has run into severe interference caused by the mine machinery. Shorty has just been flown to Butedale and has 11 fishboats to fix up. John has the plane in Simoom tonight and coming in tomorrow. Hard to keep track of them all, but they've all phoned since supper.

Cousin Rupert was starting to get a bit more spare time, and whenever he did he'd come over to our office and bullshit. The more he did the more ideas he dreamed up for us to do. Bill Peters was also very fruitful in coming up with a lot of crackpot schemes, ideas were popping up around us all day long like popcorn inside a popcorn machine. The next one we began to hear a lot about was the Stranraers. About this time there were twenty-four twin-engine Stranraer flying boats sitting at Jericho Beach in English Bay. All the way from worn out and finished to new and no time on them. Built in Montreal by Vickers in 1936, with Bristol-Pegasus engines of 1,050 horsepower each. Could take off in twelve seconds from a standing start, and they carried a two-ton load. Wonderful aeroplanes. All during the first part of the war they'd been used on coastal patrol, then the Air Force had stood them down in favour of the new Catalina and Canso flying boats.

My cousin Rupert was convinced they would make just wonderful apparatus for carrying passengers up and down the coast. A friend of his, another former Trans-Atlantic Ferry pilot, Wally Siple, had started a used aircraft business in Montreal, and he was buying and selling mainly ex-military equipment. This was before War Assets Disposal was formed, and he managed to buy all twenty-four of these Stranraers, as is/where is. He was selling them to South America. My cousin came to me and said, "Look, these fine machines, just exactly what you want, every week we're taking two more down to South America. My God, you oughta grab a couple."

Well, the logging companies were still pushing for a sched service to the Charlottes and we knew if we didn't do something pretty soon, someone else would. We had a talk to this Wally Siple and yes, he said he'd sell us a couple of Stranraers at a very friendly price—twenty-five thousand each, which for an aeroplane of that size seemed reasonable. The trouble with these Stranraers was that they were military aircraft and had never been licensed in any part of the world for carrying passengers. I took the notion rather nervously to our friend Carter Guest, and he took a very positive position on it. "Positively *not*," he roared, and thumped the table till the ashtrays jumped around. "Spilsbury, if you ever fly a passenger in a Stranraer, it'll be over my dead body!" Siple, on the other hand, with his flyboy hustler's way of looking at things, didn't seem to think this would be a problem. In fact he was so sure we could wangle a commercial passenger licence one way or another he made a proviso — no licence, no pay.

At this point it was still all a pipe dream because fifty thousand dollars might as well have been fifty million as far as ourselves being able to raise it, but I knew the logging companies were pretty serious about wanting steady air service to their camps — crew turnover was getting to be such that the boats couldn't keep up any more — so we decided to see how serious they were. If they would back a loan or give us a contract we could take to the bank, the whole thing would start to look dangerously possible. Pacific Mills at Ocean Falls, the Powell River Company, the Kelly Logging Company and the Morgan Logging Company — all big timber concerns with camps on the Queen Charlotte Islands — were the main ones who were promoting this regular service idea, and it turned out they were willing to put money into it but, being big corporations, they chose to advertise and go the formal route, inviting all interested applicants rather than just make a friendly deal with us.

A number of outfits bid. One of them was a group of Air Force veterans headed by a pilot named Art Barran, who later flew for us. They proposed using DC-3s on wheels with Norseman seaplanes distributing from Sandspit to the camps — which as we discovered later was certainly the best way to go. Head to head, landplanes are more efficient. But in those days it wasn't difficult for us to suggest to people, you know — twin engines or not, you're hanging up there over Hecate Straits, 250 miles of the roughest water on the Pacific — all that open water...in case of engine failure, in case of anything else, what sense does it make flying around there with landplanes? The Air Force had lost hundreds of landplanes up and down the coast but here were these flying boats, they never got into trouble. They could land anywhere between Vancouver and Prince Rupert. We were sold on these Stranraer flying boats and we did what we could to sell them. We made a great thing of their *seaworthiness*. People then were still very unused to flying, and were especially leery flying over water in a wheel plane.

Pacific Mills was on side, but Powell River Company kept having doubts. Twice we had the deal closed and twice they called it off. H.R. MacMillan was involved somehow, and he was having a lot to say about who got this contract. Finally they hired an expert to advise them—Squadron Leader Bill Woods. He had married into the family or something, which gave him a pretty good in. And so MacMillan and Company found a job for him and put him on to sort out this mess. An ex-Air Force type with an English accent, very fine manners—I remember he carried his handkerchief up his sleeve. We got next to him and let on that we were looking for a good ex-service officer to take over as operations manager if this deal came through. I'm sure that had nothing to do with it, but he went back to MacMillan and recommended flying boats.

Well, that did it. The companies decided in our favour. I explained the whole deal in a letter to Dad of January 19, 1946:

Pacific Mills, Powell River Company and J.R. Morgan Ltd. expressed their need for air transportation to the Queen Charlotte Islands. We undertook to provide them with a service if they would give us suitable backing. They have now at long last agreed to provide this backing and it has at last taken a legal form. A 16-page legal agreement was completed today. With the result they pay us $10,000 plus other expenses amounting to possibly another $5,000. This enables us to make a down payment of $10,000 to Siple Aircraft to give us the option of purchasing two Stranraer flying boats and to permit us to operate one for a period of 60 days as a trial. At the end of this 60 days the P.R. Co. can:

1. Cancel the deal and forfeit their $10,000.

2. Pay us another $50,000 to complete the purchase and continue the service.

If the deal continues we have four years to pay off the $60,000 mortgage and own the outfit. We start operations right away with one trip a week and they pay us $1,350 a trip. Later this will be pushed to possibly three trips a week and the price will be reduced to a point where they will be able to get their transportation at a rate that is considered competitive, probably 8 cents per mile per passenger.

Bud Lando, the lawyer I first met when I needed help collecting a radio bill from an old remittance man on Read Island back in the thirties, was working overtime hammering out the details, as he had the previous year trying to put the Grumman Goose deal together. He was a much more experienced and a much harder-nosed deal-maker than me, and I came to rely on his guidance greatly. In fact he claimed he was now having to spend two-thirds of his working day on my fool schemes and would have to give up his practice if my wheeler-dealing didn't abate. He got me in a corner and said, "Look, Spilsbury—if this goddamn deal falls through, promise me you'll go back to your radio business and let me get back to my law business?" I smiled and said, "Nope!"

What he and the company lawyers came down to in the end was they'd give us the fifty thousand dollars to complete the deal—*but* we had to make one flight to Cumshewa Inlet and back with twenty passengers and meet their approval all around before we could collect.

Now the fix was in, providing Siple could perform his magic. "Fine," he said, he was on. "If it all depends on that, the aeroplane'll do it and I'll get a licence on it." So we picked out one of the better ships, RCAF #907, Siple bent up twenty seats of aluminum tubing and expressed them out to us—and that's about all he did. Then he called in Miss Jackson. Miss Jackson was the real key to Wally Siple's system. She had two unbeatable things in her favour. First, she was an aeronautical engineer, which is the person qualified to draw up actual aircraft designs, and second, her father was the director of aircraft inspection in Ottawa, which gave her very good luck in getting anything she designed approved. As soon as we had the last seat screwed down, Siple turned the thing over to her, and she drew up a blueprint, and the next week the first Stranraer in aviation history was approved for commercial use. So hardly three weeks after my little encounter with Mr. Guest, one of these great birds flutters over the city and lands at the Vancouver seaplane base on the centre arm of the Fraser with a valid C of A, ready to load up with passengers. We changed the numbers, from Air Force #907 to CF-BYI, gave it the name *Haida Queen*, painted out the rondelles and left the rest raw aluminum. We would have liked to have painted it, but we couldn't afford to. They were big machines. They used to say it looked like the Marpole Bridge had taken off when one of them went by. I sent Dad a sketch with statistics comparing the new plane with the Waco:

Stranraer CF-BYI taxiing in the South Arm, 1946.

Docking the big beast was always a panicky manoeuvre.

Strannie interior.

Waco	**Stranraer**
Crew 1	Crew 3
Passengers 3	Passengers 20
Freight 150 lbs.	Freight 2000 lbs.
245 H.P.	2100 H.P.
Gastanks 56 gal.	Gastanks 700 gal.
Wingspan 38 ft.	Wingspan 105 ft.
Weight MT 1½ tons	Weight MT 13,840 lb.
Weight full 2 tons	Weight full 19,900 lb.
Range 350 miles	Range 1,100 miles
Speed 90	Speed 110

We notified the Powell River Company that we were ready and on March 5, 1946 they sent out about sixteen passengers including T.B. Jackson, general manager of Ocean Falls, Jim Murphy, logging superintendent of Kelly Log, and George O'Brien Sr., head of the Powell River Company—three of the most important men in BC logging—plus a bunch of flunkies and fallers and chokermen chewing snoose. I went along to see if I could keep the radio running, which was one of ours put in illegally. Bud Lando went along for the ride, Rupert was pilot, Bill Peters (who still didn't have his licence back) was co-pilot and Hank Elwyn was flight engineer. It had been the worst spring in forty years and this day looked no better.

We were terribly nervous. We had only made one flight with the Stranraer before we went into regular service. Boy I'll tell you, that was something else again. We took one flight around Vancouver— we had to test it. Somebody in their wisdom said well, we've got to take it up with a two-ton payload to satisfy Carter Guest—we've got to prove that it'll fly with a two-ton load. So we had to fill it up with passengers, and what better than get a little goodwill? We stuffed everybody in there—our secretaries, good old Norah Yates, friends of Bud Lando's, everybody we knew. We flew them all around North Van and Grouse Mountain, then came back and landed in the river. And that used up pretty well all the money we'd borrowed for fuel. We hadn't flown the aeroplane since. We couldn't afford to. We just hoped everything would work.

We left Vancouver in a southeast gale bound for Sullivan Bay, where we had arranged previously for a cache of fuel in fifty-gallon drums, and the first thing we discovered was that none of the primary instruments—turn and bank, airspeed—was working.

Immediate panic. Without them, it wouldn't be safe to continue. On the other hand, to call off the flight with all this brass aboard, and admit to malfunctioning equipment, would likely be the end of our contract hopes. I poked around, pretending to be working on the radio, but couldn't see anything obvious. Then I went back to my station and gave myself up to my nerves. No one spoke. Rupert kept going ahead. I looked out the window and found myself looking down at my old home, Savary Island, but I was too tied up inside to take much notice. Then I realized it was coming up toward me. We were losing altitude and banking. For whatever reason, Rupert had decided to land. He taxied up on the west side and shut everything down a couple hundred yards off the wharf. None of the passengers knew what was up, and neither exactly did I. Then we saw Hank Elwyn climb out onto the lower wing, take out the collapsible aluminum ladder, set it in sockets in the lower wing, climb up onto the upper wing twelve feet above and walk out along the upper wing to where the pitot head was—this is an open tube sticking out like a boat hook from the leading edge of the wing, that supplies air pressure to the instruments. Very carefully pulled over the pitot head and taped on was a French safe. When the aircraft was in storage they'd had to cover the opening to keep dirt from getting in it and, pilots being pilots, this was the first thing that came to hand. Elwyn peeled it off and bingo, the needles sprang to life.

With wind steadily rising and the sky steadily lowering, we made Sullivan Bay and took off for the Charlottes. By the time we got to Calvert Island it was very bumpy and right down-on-the-deck. Meanwhile, our passengers were rapidly becoming acquainted with some of the reasons Carter Guest didn't think the Stranraer was just the best thing for commercial flying. It wasn't lined—it was just single-skin aluminum, and it was *freezing*. I mean literally. There was no heat aboard. Our breath was freezing to the windows, there were icicles all over everything, everybody was blue with the cold. I was airsick and so was Bud Lando. There was no toilet, no facilities of any kind outside of a suction tube the crew used to use in the Force. Just a short rubber hose to the outside with a funnel on your end, and when you opened the little cock it would whistle, there was a helluva suction on it. It became quite famous later and earned the plane the nickname "The Whistling Shithouse." We used to wonder how the women made out—we put a curtain across behind the last seat and every once in a while, when there was absolutely no recourse, you'd see some woman passenger rather painfully make her way back there, draw this curtain and you'd hear this whistling sound — I guess somehow they managed, because they always came back.

Now, for all his wild talk on the ground, Rupert was very, very cautious in the air. He never got into trouble, never bent a plane — not since he went down on the farm in Sprott-Shaw's Barling, at any rate. So he told me he wasn't going to head out into Hecate Straits with the weather like this and asked me where I wanted to put down for the night.

I felt terrible of course. I was just beginning to face the fact all was lost, and found it quite hard. George O'Brien, the patriarch of the Powell River Company, took over as wagon master at this point, and directed us to Coal Harbour on Vancouver Island, where there was a wartime seaplane ramp being used by the Gibson Brothers to haul up dead whales. We started to feel a little better as the flying smoothed out on the way back to Coal Harbour, but when we landed in the middle of all this whale blubber and guts we got sick all over again. O'Brien had previously got the big Powell River Company camp at Port Hardy on the radio, and a crummy was waiting for us. We spent a comfortable night and we made the Queen Charlottes the next day without further trouble, but our overall time wasn't much better than it would have been on a good straight-through boat, and the first day had been just a dreadful ordeal. I had given up on the whole idea by this time and was particularly dreading the next leg of our flight, into Prince Rupert. We had advertised ahead of time the starting of a regular Vancouver–Prince Rupert air service, but with the logging camp deal fallen through there was no way it was going to happen. It had been arranged we pick our boss loggers up southbound two days later, however, so we had no choice but to go on.

When we arrived in Rupert there was a civic welcome to greet us. The docks were crowded with people. We couldn't believe it. The mayor and town council were there with signs — Prince Rupert Welcomes Flight Number One — they made a huge fuss of us and put us up at the hotel, all expenses paid. The March 7 *Daily News* gave us a 48-point headline — INAUGURAL FLIGHT OF VANCOU-VER–RUPERT AIR SERVICE LANDS HERE, although the story itself was suitably cautious. "The non-schedule air service during its preliminary stages will *probably* be on a weekly basis... The ship will *probably* be back next Wednesday, returning south Thursday..." The next day, when my cousin was up on the wing refuelling, they heard something up at the head of the dock, and here was a priest waving his arms and making an incantation. "For Christ sakes," hollered Rupert, who was never at a loss for the inappropriate phrase, "now they're blessing the God-damned thing!"

It needed it. Right then I wouldn't have given two bits for our chances of ever returning to Prince Rupert in that aeroplane. We got a few southbound passengers as it turned out, picked up our bosses at Cumshewa, the weather was then greatly improved, and we flew straight through to Vancouver. O'Brien was ominously silent about our contract on the way down, but as soon as he was safely on the ground in town he broke into a big smile and pumped my hand.

"If you can fly in that weather, you can fly in anything," he laughed, then gave me a cheque for fifty thousand dollars. With that we paid Siple and put in the order for our second Stranraer. We kept our word to Bill Wood and hired him as operations manager, but he proved to be a bit of a liability. One time the papers asked him what he thought about our little airline, being such a distinguished expert from outside, and he said, "Unorthodox. Yes that is what it is, *unorthodox*." This was no doubt true—if you were comparing us, say, to the Royal Air Force—but trumpeting it in the media was a poor way to build customer confidence. One paper made a headline out of it. This was so typical of him. He just couldn't get on the right wavelength. We hired two more pilots, Art Barran and Ken Wilson, added three more ground crew, including a bright young engineer named Dick Lake, and by April 1, 1946 I found myself signing paychecks for twenty-two employees.

I didn't realize how much pressure I'd been under through this Stranraer campaign until it was over. I wrote Dad that I felt kind of let down, so that it took an enormous struggle to do the slightest thing. Nothing else seemed quite important enough to warrant the effort. Glenys said I was getting too miserable to live with.

We got the big ship on a regular schedule, going up to the Charlottes on Tuesday and coming down Wednesday. On Saturday we made a trip up the lower coast to O'Brian Bay and back, stopping at Minstrel Island and waypoints. In June, 1946 we incorporated a new company for the flying side of things and turned over ownership of the planes in exchange for shares for Hep and myself, with some shares going to the logging companies in trust. After much dithering and hemming and hawing we settled on the name of Queen Charlotte Airlines Limited (QCA), which shows we were still thinking of our main purpose basically as contract fliers to the Queen Charlotte

Islands. Within months I would realize this name was much too limited, but by that time we had too much invested in stationery to change it. The notice in the British Columbia Gazette listed capitalization of the new company at $250,000. This fantastic figure was arrived at by calculating the most optimistic value of the various contracts, our goodwill, and the price it would cost to replace our old pre-war planes with brand new models of the same size. As our accountant John Weeden explained, this would allow us to have all sorts of fun ducking taxes by depreciating it all over the next few years, and of course he was right. I designed a QCA logo myself, showing a shield bearing two spread wings over a Haida war canoe, with the motto, "In the Wake of the War Canoes." In July we put on what the Prince Rupert *Daily News* called a "colourful ceremony" complete with a Metlakatla Indian "Queen" to launch our second Strannie, CF-BYL, which we christened the *Skeena Queen.*

We were getting embarrassed for space in Vancouver so we stationed the new ship at the Seal Cove seaplane base in Prince Rupert, which was still operated by the Air Force and had all kinds of unused capacity. It was much the best seaplane base anywhere on the coast, having been built early in the war at a cost of ten million dollars. During the height of its operation the RCAF had a thousand pilots and ground crew stationed there to anchor its seaplane patrols on the north coast. C.D. Howe had offered the base to the City of Prince Rupert for a dollar, and there was quite a controversy raging in the city council about whether the city could afford to maintain it. We pushed for it, of course, and so did Ernie Carswell, the new aviation sales manager for Standard Oil, who was to become one of our key supporters.

We were doing a lot of freight hauling out of Rupert for Morris Summit Gold Mines in the Salmon River valley twenty-eight miles north of Stewart, an old discovery which had been recently put into high gear. So it made sense to have BYL stationed at Seal Cove. We also used it to add a second weekly trip on each coastal route. Our weekly route schedule, as reported to the weekly board of directors meeting July 9, 1946 looked like this:

Mon.	Flight 1	Vcr to P.R. via mld 6 hrs.
Tue.	Flight 2	P.R. to Vcr via QCI 6 hrs.
Wed.	Flight 7	Vcr to P.R. via QCI 6 hrs.
Thu.	Flight 9	P.R. to Stewart)
"	Flight 10	Stewart to P.R.) 3½ hr. total

Fri.	Flight 8	P.R. to Vcr via mld 6 hrs.
Sat.	Flight 3	Vcr to Seymour Inlet)
"	Flight 4	Seymour Inlet to Vcr) 6 hrs. total
Sun.	Flight 5	Vcr to Bridge River)
"	Flight 6	Bridge River to Vcr) 3 hrs. total

TOTAL FLYING TIME 36½ HRS.

Charlie Banting told the meeting that from a maintenance standpoint the equipment could easily be flying double the time we were getting. At this same meeting we unveiled our new organizational chart. Mr. Banting as superintendent of maintenance was charged with all ground maintenance and repair work, assigning flight engineers and ground crew, recommending promotions and pay for his staff, providing fuel at the Vancouver base, keeping engine and airframe records. Mr. Hatch, although called somewhat ambiguously "executive pilot," had now been slipped into the role of de facto operations manager, pushing Bill Wood sideways into traffic. Johnny was in charge of flight crew assignments, trip routing and new fare quotations. Art Barran, as assistant to Mr. Hatch, was in charge of out-of-town fuel caches. Rupert was official test pilot. Mr. Wood, as traffic superintendent, was now in charge of reservations, ticket sales, regular quotations, passenger comfort, taxis, baggage and additional passenger insurance. In other words, he was demoted to ticket clerk. Even that turned out to be too much for him though, and eventually we slipped him out the back door. Norm Landahl, a young fellow we hired in May 1946 as bookkeeper, was to do general accountancy, banking and payroll. His job expanded so quickly we were hiring an assistant for him before the summer was out. In July we took on two more pilots, Ken McQuaig and Ken Mackenzie. In August we added Bert Toy.

The comparison between the various routes was interesting. Flying north out of Vancouver, the busiest route was the subsidized flight to the Charlottes with 231 passengers up to September 30, 1946. But the "free enterprise" run to Seymour Inlet, with 191, wasn't far behind. Prince Rupert and points north weighed in third with 103 and Queen Charlottes to Prince Rupert came last with 82. Going south it was quite different. The "regular charter" from Seymour Inlet (which included stops at Sullivan Bay, Simoom Sound and Minstrel Island along the way) led with 198 passengers. Prince Rupert and Stewart was second with 171 and QCI third with 112. Prince Rupert to Sandspit was showing 97 paid fares.

New Norseman delivers a stretcher case at seaplane ramp, Vancouver.

Johnny Hatch with passenger Doug Belyea.

Norseman picking up load at Johnson camp, Seymour Inlet.

Johnny using rope to remove overnight snow from Norseman wing.

At the beginning of August we got word of an awful uproar in Prince Rupert—Rupert had been up there with the Stranraer to fly the Stewart leg, he had got partying with our local agent, George Stanbridge, and he'd got so damn drunk he couldn't fly for two days. Stanbridge then phoned in complaining about all the embarrassment, loss of prestige etcetera which Rupert had brought upon his good work in the town and demanded we either get new pilots or find a new agent. Stanbridge was just finger-pointing, but I knew it all by heart. Rupert never changed. He'd go into these little places and get drinking and shouting and insulting everybody to the quick with his damned Limey flyboy arrogance and he would make people just hate us. It was embarrassing to the company and it was embarrassing to me personally, because I felt a family responsibility for him.

I suspended him. But I'll say this for Rupert, he took it like a man. He admitted he couldn't go on like that and offered to help out in some other capacity. We kept him on with the official title of company test pilot. But the drinking kept on, his and everybody else's. It seemed everybody in the aviation business except me was devoted to hard drinking. It was part of the life. I would have liked to do something but I didn't know where to start. I despaired at our slowness to acquire discipline and a businesslike manner befitting our new-found position of eminence on the local flying scene. Rupert wasn't the only one who'd scare the hell out of passengers by bragging about near-misses or God-damning this haywire outfit and its haywire equipment. At the June 1, 1946 board meeting we passed a resolution requesting "that all personnel be cautioned to make no derogatory remarks, jokingly or otherwise, in the presence of passengers..."

It didn't do any good.

The good news was that in spite of our spiralling expenses and swelling payroll the flying operation showed a six-thousand-dollar profit over the second quarter of 1946.

We were getting a lot of charters that were too big for the Waco and too small for the Stranraers, so we decided it was time to get a Noorduyn Norseman. The de Havilland Beaver was a year from production, so the Norseman was still considered the best bush plane you could buy. It was first built in 1936 by Robert Noorduyn, a Dutch-English engineer who had designed bush planes for Fokker and Bellanca before emigrating to Montreal in 1935. The Norseman was his only solo effort and it was specially designed for use in the

Canadian North. It had a high wing for stable flying and easy parking near docks and shores, a large tail for good control on the water and oversize removable doors for bulky freight. It had a wingspan of 51½ feet, used a 550-horsepower Pratt and Whitney Wasp motor and carried nine passengers. The Norseman had a legendary career in the northern bush and also distinguished itself during the war, particularly in the US Air Force. The trouble was, being the prime work plane of the day, it was hard to get and came pretty high. From the manufacturer, Canadian Car and Foundry in Montreal, they ran about forty-five thousand dollars, and good used ones were rare. By 1946 there were a few war surplus models coming available but you never heard about one until it was gone.

On May 22 I wrote direct to A.E. McMaster, general manager of the War Assets Corporation in Montreal, who sometimes took his holidays on Savary Island. I complained that we had repeatedly requested craft from the local War Assets office and was told no Norsemen were available, only to find out later that five had been sold to Yukon Southern Air Transport and three to CPAL. "Our need for these ships is imperative," I told McMaster, "and in desperation we are writing you direct to ask if you are able in any manner to facilitate the release of at least one such Norseman, making it possible for us to handle a great deal of mining charter work which is now available, and which must be handled before the lakes again freeze up." McMaster took the beef seriously and put me in touch with his chief of aircraft sales, J.R. Douglas. We bombarded Douglas with telegrams and letters for a couple of months and finally in July he offered us an old pre-war Mark IV Norseman without floats for fifteen thousand dollars. It was located in Trenton, Ontario. We scraped together five thousand cash of our own, touched the bank for a mortgage on the balance and bought it sight unseen. Johnny Hatch had a younger brother George who was also a pilot, and he'd been looking for an excuse to bring him out from Toronto, so we sent George up to Trenton to take possession of this Norseman. He ferried it out for us the very day the papers cleared, arriving in late August. It went into service soon after and seldom rested after that date, except for refit.

Norsemen became easier to get as the Beaver came into use and displaced them, and we gradually picked up more over the years. At one time we were flying fourteen at once. I think we must have had the largest Norseman fleet ever assembled in civilian times. Some were great, some were terrible, most were adequate. The wartime models, the Mark VIs, were built often with substandard materials and tended to be heavy, with low performance. But that first one CF-EJB, turned out to be one of the better ones.

During the summer of 1946 we became desperate for a home for our flying operations in Vancouver and I waged another bureaucratic campaign, this time to obtain some office and hangar space at the Vancouver Airport on Sea Island. We got some hangar and office space from TCA for awhile, but the boys caught a Stranraer on fire one day while they were welding up an exhaust ring and TCA threw us out on our ear. Then, after making appeals to half the MPs in Ottawa we got to C.D. Howe and were allowed to rent part of the old Boeing plant. This provided us with a big haul-out ramp complete with electric winch house and a small administration building. We stayed there until we finally acquired the old CPAL overhaul premises, and we set up a proper terminal and got down to business. Between April and September of 1946 we carried 2,369 passengers, hauled 96,463 pounds of freight, flew a total of 146,588 miles and employed thirty-three people. Total operating revenue was a shade under $130,000.

Mercy Flights

IT WAS A LITTLE TOO GOOD TO BE TRUE, but we were about to be brought down to earth with a bang.

As the only planes available to most parts of the coast, we did more than our share of mercy flights. It was always the same story. Old Alan Greene or some harried logging-camp timekeeper would phone me up with a logger dying at Rock Bay or Pender Harbour or Alert Bay, and it would usually be the worst flying conditions. Zero/zero visibility, seventy-knot gales or no daylight. But it was a time when what you did could make the difference between life and death. The first few times you felt it was worth risking your entire operation, your entire life, and when it came off you felt wonderful; all your work was vindicated. Pilots couldn't resist the challenge. It gave them the opportunity to break all the normal flying rules and be written up as heroes on the front pages of the papers. As owner, I couldn't resist the fabulous free publicity.

"MERCY PLANE FLIES INJURED LOGGER HERE... Thwarted three times by sweeping rains, fog and wind, a seaplane making a flight to Rock Bay to bring a seriously injured logger here, succeeded Sunday... DIES IN PLANE... Pilot Len Milne, flying a Spilsbury and Hepburn plane, lost a race against death through chilly BC skies today in a mercy flight attempt to save the life of an injured Simoom Sound man... PLANE PICKS UP INJURED BC MAN FROM SPEEDBOAT... Flagged down in Hecate Strait Thursday afternoon by a circling speedboat, a 15-passenger flying boat operated by Spilsbury and Hepburn turned back to Prince

Rupert to rush a critically injured logger to hospital... Gus Peterson, about 45, was injured at Kelley Log, Cumshewa, when a logging axe he was using to limb a tree snagged and struck him... Setting the Stranraer down cautiously in the heavy seas, pilot W.J. (Bill) Peters arranged a rendezvous in a sheltered cove... MERCY FLIGHT BRINGS LOGGER TO HOSPITAL... QCA brought an injured logger to Vancouver Tuesday night from Fraser Bay, 120 miles north of here. The injured man, 26-year-old Charles Bayne, is a high rigger employed by Maloney Logging Co. He fell 120 feet from a tree, sustaining back injuries... ICE BLAST VICTIM FLOWN TO CITY... Harold Stewart, 40, suffered face, leg and hand injuries, when after setting off the fuse in a frozen log pool and running, he tripped... Ken McQuaig, pilot for QCA, brought Stewart here... LEG TORN OFF, ENGINEER DIES... Almond Maughan, 35-year-old engineer on the Edmunds and Walker Ltd. fish packer *Quathiaski 14*, died late Wednesday in Gardner Canal after his leg became caught in a propeller shaft. QCA flying boat rushed to the scene... was forced to land at Powell River when fog closed in... TIMBER LOADER FLOWN TO HOSPITAL... D.A. Smith, loader for Universal Timber Products, is recovering in St. Paul's Hospital today from injuries received Thursday afternoon when logs rolled on his chest... Smith was flown to Vancouver in a Spilsbury and Hepburn seaplane piloted by Art Barran... MINER FIGHTS FOR LIFE AFTER MERCY FLIGHT... Walter Wuori was struck by a two-and-one-half-ton steam shovel... brought here by Queen Charlotte Airlines pilot John Hatch after risky flight through snow storms... lakes all frozen over, pilot Hatch landed floatplane *Nimpkish Queen* on snow at Bridge River... FLIER DEFIES DISTANCE, SNOW, TO REACH MOTHER..."

We began keeping a book on all the mercy flights and by September 1946 we informed the Prince Rupert *Daily News* that we had flown one hundred of them in five months. It was getting to be such a regular thing and such a nuisance that it affected morale, both of crew and customers. Our pilots were so constantly thrust together in close quarters with broken, burned, bleeding and dying bodies they began feeling like medics at a particularly active battlefront, and terrified customers found themselves sharing cabin space with battered corpses far more often than we wanted. But what were we to do? As the only available air service in a large territory whose residents engaged in some of the most unsafe work in the world, this was the role that fell to us. But it was costing us a lot of money, and at a directors' meeting in September we rather reluctantly took the step of setting a special rate for stretcher cases, equal to double the normal fare. They usually took up at least two seats, and meant a lot of inconvenient scheduling besides. At the same time it was resolved that the company discourage wherever possible the carriage of corpses, but that when same necessary double fare be charged.

Prince Rupert was proud of its newly acquired seaplane base, proud to be the home of the *Skeena Queen* and happy with us for bringing it into the air age. On July 23, 1946 the *News* celebrated the new era in an editorial:

> A novel ceremony at the Seal Cove air base yesterday afternoon when the christening took place of the *Skeena Queen*, second of the air liners to be placed in service between Prince Rupert and Vancouver, served to accentuate the fact that the pioneering of Queen Charlotte Airlines in commercial aviation from here has, in its so far brief span, done much to make Prince Rupert and its people air-conscious. We have long looked for the day when such a regular service would be available here. Now it is an accomplished fact. All that is needed now is a reduction of the rate tariff to a more reasonable figure to make air travel between Prince Rupert and Vancouver as well as other points on the coast an accepted thing which will demand a much more frequent service than the two regular flights a week which have now been instituted.

The two Prince Rupert papers, the *Daily News* and the *Evening Empire*, treated our operation as front-page stuff, writing up every mercy flight and even publishing passenger lists of arriving and departing flights. They were grateful. They seemed to feel we were flying in there as a personal favour to the people of the area. We kept a crew up there, living in one of the hotels, and we felt like family.

Our first Stranraer, CF-BYI, was used in its original colour — aluminum. We had neither the time nor money to spend on paint.

When we got the Powell River Company contract and bought the second one, the *Skeena Queen*, we were a bit more flush so we gave it a complete paint job. The hull (fuselage) was painted black. The wings (all four) were painted Loening yellow, Loening being a well-known manufacturer of aeroplanes who used a special bright yellow. Our idea was to copy the Waco and the Waco was in these colours. I suppose someone thought it would give us a company look. So nothing would do but we paint this great long Stranraer hull *shiny black*. It was the most godawful thing—they called it the flying coffin.

On August 31 the crew were having supper in a cafe when a radio message came in from Stewart—a little girl with a bowel obstruction was in critical condition and could we fly the mother and child down to Prince Rupert for an emergency operation. Our pilot, Ken Wilson, just three months with the company, decided to make the flight, even though it was an hour and a half after grounding time. Just two days earlier he'd made a mercy flight to rescue the Stewart mine manager's wife, so it was routine to him. He knew it would be dark coming back, but it was clear and he'd be able to land by the lights of the town reflecting on the water. The co-pilot was George Hatch, Johnny's younger brother, just on his first week with us after ferrying the Norseman CF-EJB out from Toronto. They paid their bill and went out to the aircraft. The flight engineer Jens Madsen went with them, as well as Lloyd Douglas, the engineer of the Waco, which happened to be in town. A bystander named Nick Killas, who managed the restaurant, also decided to go along for the ride.

The flight took off at 7:07 P.M. and arrived in Stewart in just under one hour. The 23-year-old mother and her 3½-year-old daughter were taken aboard and BYL headed right back to Prince Rupert, but thick fog had rolled into Rupert harbour. The plane was equipped with radio but limited to DoT "tower" frequencies for clearing in and out of government airports. We badly needed our own company frequency but so far Ottawa would not assign us one. Our Prince Rupert agent, George Stanbridge, couldn't make contact to warn about the fog, and the aircraft didn't report its position to him. We had written so many letters about this radio problem, we tried so hard, we said we have the ability, we have the knowledge, we have the equipment, we need communication with these aeroplanes—but the DoT wouldn't listen. So we had to bootleg calls on the busy ship-to-shore frequencies, and it was very hit-and-miss. Our operations manager was constantly after us to improve radio communications, claiming the present system failed "nine times out of ten," but until the DoT moved there was little we could do. At 8:50 P.M. they heard the plane return and circle over the city above the fog. The DoT wireless station on Digby Island tried desperately to make radio contact, but BYL remained eerily silent. A US Navy cruiser that was in the harbour, the *Tucson*, plied the murk with its twenty-four-inch searchlights, hoping to guide the plane down, but got no takers. Finally the noise of the plane faded away. We thought she would put down somewhere else, perhaps Port Simpson, where there was a hospital, but we knew she would call in immediately if she landed anywhere near a working phone. After that we all just waited — and waited, and waited. But the night passed with no word from the *Skeena Queen*.

Next morning John Hatch and I caught the morning BYI flight to Rupert and started a full-scale search with boats and aircraft. They could have landed in some isolated bay where it wasn't possible to get word out. They had five hours' fuel aboard. US Coast Guard planes from Ketchikan and Juneau came down to help. The RCAF sent planes up from Comox. The little Waco flew from dawn till dusk — day after day — searching mountain sides when the weather permitted and all areas of the inlet for floating debris. Nothing — many reports of smoke and flares — all false as usual.

It was still quite possible that the plane had somehow been damaged or even sunk on landing, but that the occupants were ashore somewhere in the trackless waste of Portland Canal trying to get our attention. Then on September 2 a fisherman searching off Finlayson Island in Chatham Sound, eighteen miles north of Prince Rupert, found a badly-battered body floating face up in the water. It was identified as that of Mrs. Margaret Dempsey, the child's mother. It was pretty clear she'd been through a violent accident. Searchers in the same general area also recovered three lifejackets and an oil tank from the strongest structural part of the plane. We knew it couldn't have come loose unless the plane had been pretty thoroughly smashed apart. The next day the Reverend Basil S. Prockter found Ken Wilson's body beached at the high-tide mark while searching the shore of the Tsimpsean Peninsula south of Finlayson Island. Later Tuesday another searcher found the body of the little girl floating southwest of Finlayson Island. All finds were in the same general area slightly below Port Simpson, ten or so miles northwest of Prince Rupert. We searched the bottom for days with dragging equipment, the naval ship HMCS *Charlottetown* combed it

The "Flying Coffin"—Stranraer CF-BYL.

Mercy flights beyond counting. Charlie Banting looks grimly on as another stretcher case comes in after dark.

The Skeena Queen, *last seen circling Prince Rupert on August 31, 1946.*

with asdic and 250 people from town searched every inch of the shoreline, but we found nothing more.

What happened we never knew. Two eyewitnesses turned up who'd been camping on top of a mountain outside Prince Rupert, above the fog, and actually saw BYL circle twice over the city at about 9 P.M. then set a course to the north. They said the USS *Tucson*'s searchlight hadn't shown up at all from above. From there, a timber cruiser in Wark Canal saw the plane heading northwest at 9:10. A third witness saw the plane off Port Simpson at 9:12. The sky over Port Simpson was clear with visibility of twenty-two miles, although by this time of course it was quite dark. Many people in the village heard the plane. Ken Wilson's watch had stopped at 9:23 P.M. We could only assume he flew into the water in the dark or tried to make a landing and hit something—a log or one of the area's many reefs.

Half the population of Prince Rupert turned out to search—service clubs organized an armada of rowboats to comb the shoreline and the ladies' auxiliaries provided a mountain of sandwiches. Mrs. Dempsey came from one of the area's pioneer families, and the restaurant man, Killas, was well known. It really shook the town, and it really shook us. Only a year in business and our first loss—seven lives—we couldn't believe it. It was an especially terrible blow to John Hatch, who was very fond of his younger brother, and had worked hard to bring him out from Toronto for this job.

It taught us a lesson—a bitter lesson: never let the urgency of a "mercy flight" affect your normal judgment concerning safe flying. From that day on our company Operations Manual dealt very severely with the matter of minimum visibility and weather conditions, and the importance of not allowing the emotional demands of "mercy" flights to influence a pilot's decision. I wrote Dad on September 15:

> There is no doubt the trip should not have been undertaken and would not have received authority from our office had the pilot requested it. He was one of the oldest we had with most flying time, but even so had been kept as a co-pilot for over three months as we were not satisfied with his work. This was his first trip as captain and then only because we were sending the plane out on three weeks' straight

freight hauling where we thought he would do alright. It was distressing to find all three widows were without adequate insurance protection. Our group insurance and pension plan was at that time before the board and has since gone into effect. Also we were in process of increasing our aircraft insurance by $15,000. However at that we did not get off too badly. Here are some interesting observations:

—After the crash passenger traffic on the airline showed a marked increase. The publicity we received through the accident could not have been bought with any amount of money. Thousands of people know of QCA now that had never heard of it before.

—Lloyd's insurance on the balance of our aircraft has since the accident dropped from ten-and-a-half percent to eight-and-a-half percent. Apparently the reasoning is that we will now be much more careful than previously, and they are undoubtedly right.

—The Department of Transport has not called for an investigation and it is very doubtful they will. We don't know what effect this will have on our application for the licence to make scheduled flights into Prince Rupert.

While the whole affair was of course highly regrettable it is nevertheless one of the things we must face from time to time, and it is generally felt if we can ride over this one we can handle anything that comes in the future.

One good thing came out of it. The DoT began moving on a company radio frequency assignment without further delay. Almost the next day.

And we never again painted one of our Stranraers black!

Of course we now found ourselves in a proper bind for planes. Luckily, our Norseman EJB was just waiting for its final registration papers to come over from Carter Guest and went in service on September 16, 1946. But we got even luckier. BYL had been in

Prince Rupert to rush six hundred tons of freight from Rupert and Stewart up over the mountain into the Morris Summit Goldmine before the water on Summit Lake froze up for the winter. It was a good contract we'd been counting on to pay down our bank loan and create a little breathing room financially. We were carrying it on as best we could with the Norseman and BYI when we could spare her for a few hours, but it was touch and go. In fact we might have got the job done, but I hit on the idea of putting the bite on old Colonel Ralston, the mine owner, to loan us twenty thousand dollars to go back to Siple for a replacement Stranraer. The argument was that this was the only way they'd get their mine supplies in for the winter. They agreed, and our third Stranraer, CF-BYJ, was delivered to us before the end of September. At the same time the insurance company delivered us a cheque for $22,500 for BYL. This was technically the timber consortium's money as we'd had to take the insurance out in their names, but since it was intended to be invested in a Stranraer for the next several years, we figured they ought to be willing to put it back into a third plane. There was a good Stranraer with new engines for sale in Montreal for the bargain price of $12,500 and we set Bud Lando after them to get that.

Incidentally, Summit Lake was just over a mile long and had free icebergs floating around in it. The wind had to be blowing a certain way to move the bergs down to one end before we could land, and we'd always make a pass over the water to take a look before we put down. The bulk of the lake was caused by a glacial plug, which at a certain time of the year would melt and send half the water sluicing down the mountainside. The men used to have a bet on the day it would happen. This worried us because we were landing and taking off every day and we didn't want to be in there when the plug went.

We never did get caught, but there were two old prospectors tunnelling away in there. The road ended on the south side of the lake and they had a raft they would paddle across with their supplies. They went down to Stewart or to Hyder, Alaska as the case may be, Stewart for supplies and Hyder to get drunk—it was like one town with the international border running through the middle. People just wandered back and forth, but mostly they wandered to the Hyder side because it had the tavern. Anyway, these two go back up to the lake with a load of supplies and record-breaking hangovers and they're painfully paddling their raft back home when the ice plug lets go. They haven't heard about this phenomenon and it's not immediately apparent to them because they are up at the other end.

What they do see is a big mass of rock slowly rising up out of the water in the middle of the lake. The whole landscape appears to be transforming around them as the shore rises up out of the water. They're completely spooked. They hightail it for the shore and one of them runs all the way back to Hyder raving like a madman. He didn't know what it was but he wasn't sticking around to find out.

We had been running a de facto scheduled service into Prince Rupert since March, although every flight had to be disguised as a charter and we were unable to advertise fares and times. We instead ran ads, in all our regular stops — Comox, Powell River, Prince Rupert — offering something we called "regular charter." Instead of mentioning times or fares we advised readers to phone our local agents — George Stanbridge in Rupert, Al Kerr in Comox, Allan Morrow in Sandspit. We rather expected to hear from Carter Guest about it, but we figured if we got people hooked on our service the ATB would have a hell of a time chopping it off later.

On September 4, 1946 we screwed up our courage and had Esmond Lando draw up the papers and apply for a licence to fly a scheduled service Vancouver–Prince Rupert. This was quite the most serious hearing we'd been in for yet, and again we were opposed by CPAL. The ATB didn't take much notice until December, when CPAL made separate application for a scheduled licence into Rupert. We countered by applying for scheduled licences on the CPAL routes to Powell River, Alert Bay, Comox, Tahsis and Zeballos. This was the heart of the CPAL empire, and we had no expectation of getting any of it, but Lando thought it would be good tactics. The ATB, consisting of acting chairman Alan Ferrier, Romeo Vachon and C.S. Booth, finally came out to Vancouver to hear our arguments on February 25 and 26. We flew people from all over the coast in to testify QCA was giving excellent service, the first they'd ever had, they wanted them to run the Prince Rupert service. The CPAL was taking the hearing in deadly earnest because, if you know anything of Grant McConachie's early plans for CPAL, it was to make it the first trans-Pacific air carrier to use land planes by taking short hops over the polar route to the Orient, and Vancouver–Prince Rupert was to be the first hop. However, we had a very well-prepared application, we had very good grounds because we had been providing service up the coast where CPAL hadn't bothered and we had tremendous public support. Grant McConachie was trying to talk up a Prince Rupert airstrip on Tugwell Island if he got the licence, but it was difficult for him to be very convincing because of CPAL's poor record up to that time. He was also bucking the political tide in Ottawa, where the all-powerful C.D. Howe, through his sponsorship of TCA, had become CPAL's enemy, eager to keep the private airline down. This made the most powerful man in Ottawa very supportive of our coastal ambitions.

After the first day of the hearing it looked very rosy for us and very bleak for CPAL. When the court adjourned for the day, good old Grant McConachie came waltzing over to me with his great beaming smile, threw his arm over my shoulder and said, "Jim, let's you and me have a talk." Well, he talked to me like a Dutch uncle and I'm glad he did. He said, "Look, we are in for big things, we need this, we need it badly, we have the aircraft to carry it out. You're doing a darn good job on the lower coast where you've gone, but you haven't got the kind of money to get the aircraft you need for Prince Rupert, and you can't get it. You're out of your class. But I tell you what you do. You withdraw your Vancouver–Prince Rupert application and I will give you all our coast licences, Vancouver–Powell River, Vancouver–Nanaimo, Vancouver–Alert Bay, Vancouver–Tofino, Vancouver–Zeballos and Ocean Falls — everything." He was offering us the whole coast.

Grant McConachie was four years my junior, but as a wheeler-dealer he had me backed off the map. I had to sit down and think about this. Certainly what McConachie was offering seemed like a wonderful deal. The south coast was where all the people were. It was a compact flying area we could handle nicely doing the kind of short-range seaplane flying we were already pretty good at, with the kind of Norsemen and Stranraer equipment we could afford to buy. It would give us a formidable beach-head in the flying world. McConachie, who had been flying short-haul for years in the north, was gambling that there would be more profit in flying long-range to places like Sandspit and Rupert that had less population but for whom the aeroplane was the more important way out. In our own experience Rupert and the Queen Charlottes had certainly produced the most business, and under McConachie's deal we would have to let them go. What about all these north coast people we had imposed upon to come down and testify that we were the best company? To the people of Rupert, we were their own airline. If I turned around and delivered them up to a bunch of CPAL strangers, what would they think? It was a hell of a thing to have to decide over supper. The future of the company was at stake. Hell, the flying map of Canada was at stake. I lay awake all night. Damn McConachie and his deals!

In the morning I decided to accept.

So the thing was, shall we say, settled out of court? We withdrew our application for Prince Rupert and CPAL withdrew their opposition to our applications for all the places on the south coast that we only listed for the nuisance value and never expected to get. The ATB was left with nothing to decide, which irritated them no end. They held up approval of the new licences until May and in the meantime forbade us from continuing our "regular charter" into Rupert, which left the city without air service and us without a big part of our income for three months. The same on our regular runs into Powell River, Seymour Inlet and Sullivan Bay. We had to lay off fourteen men and by the time the ATB finally got around to completing their miserable paperwork they'd almost succeeded in forcing us into bankruptcy. We'd burned up all the working capital we'd amassed from Summit Lake and gone through the windfall money from the insurance on BYL as well. But we'd managed to last them out, and now nothing stood in our way.

QCA was now a major scheduled airline. By August the following year we were showing a twenty-eight-thousand-dollar surplus and I was able to report to our directors that while some winter slowdown could be expected, the situation was not parallel to 1946 because we were now flying two trips daily to Comox, there was a non-scheduled trip to the west coast, there was a big increase in traffic on our run to Sullivan Bay and volume was expected to be much larger than in the previous year. I wasn't sticking my neck out much with that guess. When the Dominion Bureau of Statistics published its figures for the year 1947, we were confirmed as the fourth largest airline in Canada, with only TCA, CPAL and MCA (Maritime Central) having flown more revenue miles. It was time to put our days of seat-of-the-pants flying behind us.

SEVEN

Bushplanes with Barnacles

LOOKING BACK, the radio business was the best possible preparation for the move into aviation. Almost every camp we went into, almost every mine we went into, there was a Spilsbury and Hepburn radio there and I'd been up before—by boat, by packhorse—God knows, any way you could get up there. And then with the little Waco I'd get in and I knew all these people. Gee, when we started hauling passengers for them we really had the public on our side. Wherever we went up the coast, wherever I travelled up there I'd land among friends. Every time. Slap me on the back—how ya doin? When ya goina get some better aeroplanes? Gosh we like this service!

When we got our radio frequency in the aftermath of the BYL crash, we had communication working for us. From then on we put Herb Hope's little AD-10 sets in every port of call, every little place where we might only stop twice a week. When they had passengers they'd get on the air, they'd talk to Vancouver, they'd talk to the aeroplane, the aeroplanes would talk to them, and that built the business just like you wouldn't believe. So here, the radio business came to the assistance of the flying and made the whole thing possible. Nobody else including CPAL ever had radio communication anything approaching that. The Powell River Company contract and the radio, right from the start, was what gave us the lead over everybody else trying to do the same thing.

Once we realized that we were now in the aviation business—I wouldn't say we were out of the radio business by any means; it was still there but I left it mostly to Hep and Jack Tindall because the aviation business obviously had much greater potential—we started to get all sorts of ideas. The pilots that we were hiring came up with some—heck, Chamiss Bay needs a service, and Stewart, they need flying up there, nobody's doing it—there were just all kinds of ideas. We went out and got some more aeroplanes and experimented with it and it just took off like wildfire.

CPAL had been quite willing to give up on the coast routes because they considered them money-losers. And in their experience no doubt they were. But in the post-war period the coast north of Vancouver became active to a degree that is hard to realize now. Almost every little inlet and cove up there had a camp in it going full bore. There were a lot of people came into the centres like Minstrel Island. There were a lot of camps would use Minstrel Island as the headquarters, and when we put a Stranraer in there on a daily basis we'd pick up a load. In fact we'd frequently have to run second sections to carry the traffic.

Those were our mainstays, places like that. The revenue-producing ones would be Minstrel Island, Alert Bay and, on the west coast, Tahsis and Zeballos. We caught just the tail end of the mining boom at Zeballos, but it was mostly logging. Gold River was opening up and the East Asiatic Company was going in and getting big. Oh, we had a lot of other places up the various inlets where we landed. Take Glendale Cove. For a year or so a place like that would be very busy, with just one large logging camp getting set up. When you go up the coast now all you can see are scars on the hillside where the camps used to be, and there's nobody around. Centres like Minstrel Island, there's almost nothing there now. We were there when they were all booming and we couldn't help but boom along with them.

We opened new routes like Vancouver-Comox. It was a natural. Vancouver-Powell River—that was an epic, getting that airport built at Powell River. I'd go up there and attend meetings about twice a month. Airport committee meetings. With the Aero Club of Powell River. And the community. And representatives and so on—we were trying to get money out of everybody. To get on with the building of this thing. I spent many, many hours lobbying DoT officials in Ottawa over Powell River. It took years. I think it was 1954 before it was finished to the full thirty-two-hundred-foot length and we were able to land on it with DC-3s.

We also had Vancouver–Nanaimo, which we inherited from McConachie, who *knew* it couldn't be made to pay. We found out it couldn't, but we were stubborn, we held onto it for a long time. I think our fare over there was $4.60. There was an excellent ferry service with the CPR, running many times a day, and we were competing with that. Hopeless. Especially being stuck out at Cassidy airport, it wasn't convenient. We did want to start harbour-to-harbour — Vancouver Harbour to Nanaimo Harbour and Vancouver Harbour to Victoria Harbour, but we could never but *never* get permission to fly an aeroplane into Vancouver Harbour. Carter Guest was dead against it. He would always refer back, oh, twenty years before, where some pilot had hit a tug's towline. This guy was taking off in Vancouver Harbour, got tangled up with a tug towing a barge and doused a bunch of people. I don't think anybody was hurt but he wrecked the aeroplane, so Carter Guest made up his mind that Vancouver Harbour was *not* for flying. We wanted to fly in with Stranraers. We applied over and over again but couldn't get to first base. So it's quite interesting for me now to see modern airlines doing that very thing, with twenty-passenger Twin Otters and many times the congestion we had in the forties.

We were not permitted to fly Vancouver to Victoria because that was a TCA route. We had to fly via Nanaimo. We had Ansons flying from Vancouver to Nanaimo, up to Comox, and from Comox, Nanaimo, Victoria. We had a scheduled service from Comox to Victoria, stopping at Nanaimo and connecting to Vancouver. We tried hard to make that pay off but it was no good.

Our original contract with the logging companies to fly from Vancouver to the Queen Charlotte Islands wound up as soon as Canadian Pacific Airlines inaugurated their scheduled service flying Vancouver, Port Hardy and Sandspit using first of all Lockheed Lodestars, then DC-3s, with amphibious PBYs (Cansos) running a shuttle between Sandspit and Prince Rupert. The new CPAL service, however, did not provide for any distribution to other points on the Islands including Masset, Port Clements, Skidegate, and Cumshewa, so it was agreed that QCA would continue to provide this service, as well as local traffic between the Islands and Prince Rupert and between Prince Rupert and Stewart, Anyox, Arrandale, Port Simpson, etcetera. The Stranraer flying boats were too large for this purpose, so we used Norseman seaplanes, which carried six passengers. We maintained fully equipped bases at both Masset and Prince Rupert. Then we ran into trouble with Department of

Transport, which refused to approve single-engined aircraft for use across Hecate Strait. In an attempt to overcome this obstacle we arranged to purchase two old, tired De Havilland "Dragon Rapides," CF-AYE and CF-BND, from Central Northern in Winnipeg.

The Rapide was a wood-frame-and-fabric job, a biplane with two "Gypsy Queen" engines mounted in the lower wings. They were used all over the world, but mostly as land planes, in which configuration they had fairly good performance. On floats it was a very different story. With a full load they would not maintain altitude on one engine, but would merely extend the glide. Since the Gypsy Queen engines did not have the reliability of an American Pratt and Whitney Wasp, as used in the Norseman, we argued that the Norseman with its single engine was in fact a safer aircraft than the twin-engine Rapide, but the argument did not impress the DoT. The Rapide was an *approved* aircraft in their books, and that was that. We dispatched both Rapides to the Prince Rupert–QCI service and hoped for the best, but they proved very costly to maintain and they didn't make us any money. We lived in dread of the first engine failure, and just hoped it wouldn't occur in the middle of Hecate Strait. Then one day it happened, but not the way we expected.

Roy Berryman was our Rapide pilot, stationed in Prince Rupert. This particular day, July 29, 1949, he had three lady passengers and some air freight for the islands. He was all fuelled up and loaded to go when the local weather at Prince Rupert deteriorated to the point that he was obliged to cancel for the day, so he sent his passengers into the Hotel Prince Rupert for the night, hoping for improved visibility in the morning. As he had hoped, the weather was much improved, so after an early breakfast he got his three passengers aboard and took off. The freight was already loaded, and he had topped the tanks off the night before, so he got away without undue delay. He was at four hundred feet above Digby Island when his starboard engine quit. He was not unduly concerned and started into a turn to bring him back for a landing in Seal Cove on one engine. Quite routine—so far. But then the port engine conked out and he had nothing. He was right over the middle of the island, with no hope of gliding to the water, so he picked out a couple of likely looking pine trees and plunked her down between them for a soft but messy landing in deep muskeg. The trees took all four wings off, and the engines. This is the standard bushpilot manoeuvre, to try and get rid of the wings first off because that's where the fuel is and its very

Dragon Rapide CF-BND. You'd think we'd know a plane that looked like this wouldn't fly.

The Rapide's Gypsy Queen engine kept us all guessing.

Arrival Digby Island, July 29, 1949.

Pilot Roy Berryman takes leave of his command.

prone to blow up, although in this case it didn't—for a very good reason I will go into presently. The pontoons were torn right off by the struts. All he had left was the cabin, which was reasonably intact, and nobody hurt. He was about three miles from the shoreline, however, and no one had seen him go in.

When he didn't arrive in the islands, and hadn't returned to base, the worst was feared and all available aircraft went out over Hecate Strait to search. Vancouver was immediately notified, and Johnny Hatch and I took off in a Norseman to assist. By the time we got there he had been located. He had wisely stayed with his plane and when the CPAL Canso was coming in on the sched flight from Sandspit, the pilot spotted him. Roy had removed the landing light from the nose of the Rapide, found some wire and hooked it up to the battery, and caught their attention by flashing an SOS. All in all, he handled everything just perfectly through the whole episode.

Digby Island is covered with a heavy growth of scrub pine and underbrush and muskeg swamps, and it took several hours to make our way in to the site of the crash. The three ladies were escorted back to the Hotel Prince Rupert, and their baggage was delivered to them. At this time I do not recollect if, or how, we eventually got them to their destination, we were just so thankful no one was hurt.

The underlying cause of the failure was disclosed by the local police. An Indian fisherman from nearby Metlakatla village had been boasting to some of his friends about how well his boat ran on high-octane gas. The night before, he had quietly tied alongside the Rapide and siphoned all but a couple of gallons of gas out of her tanks, which were conveniently located in the lower wings and just the right height to siphon. I don't think they ever got enough hard evidence to apprehend him, but he was generally known as a bad actor. Here's where Carter Guest had a point with his inside fuel gauges, because on the Rapide the pilot always had to dip the tanks with a measuring stick. But since Roy had done this the night before it never occurred to him that he should do it again next morning, especially with a watchman on duty.

We went back to the site first thing next morning to bring out the rest of the air freight, which contained, among other things, three cases of hard liquor for the hotel at Skidegate. When we got there the three cases of liquor had vanished. The only thing left in the way of freight was three large "dry packs" of ice cream. The ice cream had by now gone soft but the containers had some value, so I undertook to empty them onto the muskeg.

You have to picture it. The muskeg was very deep and very soft. To get around you had to lift your leg way up, like wading in powder snow. When I tried to lift this heavy container full of ice cream, I sank down to my waist. I had to hold it over my head, and when the big gob of gooey ice-cream came unstuck, it nosed around on the moss like a friendly seal, then turned and tried to slither down into the hole with me. I realized the absurdity of the situation — we had dropped a plane worth many thousands of dollars and narrowly avoided losing four lives, and here I was making an ass of myself over some containers not worth much more than the change in your pocket. But once launched on the project I couldn't seem to stop myself. Four gallons of vanilla, four gallons of chocolate, and four gallons of strawberry — I ended up in to my armpits, swimming in an enormous three-flavour sundae.

The engineers managed to salvage some of the instruments from the cockpit, but when they went back in to strip the engines, we were amazed to learn both of them had vanished, complete. How the culprits ever managed to extract those two engines out through that ocean of mush, I could never imagine. They probably weighed seven or eight hundred pounds each. It must have taken incredible determination, and to what end? I kept my ears perked up for rumours of a Metlakatla gillnetter with twin six-cylinder Gypsy Queen aircraft engines in it, but never came across a single clue. I didn't want them back, it's just that this was an upside-down engine, with the six cylinders underneath and the crankshaft on top, and it kept me up nights trying to figure out how those dedicated pilferers would plumb it up in a boat.

That took care of Rapide CF-BND. Shortly after that we made a deal with the Powell River Company to lease one of the Grumman Geese I would have bought in 1945 if R.M. Andrews Junior hadn't decided they were obsolete. We put it on the northern route and brought Dragon Rapide CF-AYE down to Vancouver, where we removed the engines and sold them. The airframe we presented to the Vancouver Airport Fire Department, who took it out to centre field and set it on fire to try out their new high-pressure fog equipment.

Johnny Hatch had broken in as a greasemonkey on Rapide AYE back in 1934 in Ontario and refused to be a party to this ignominious ending, but I must confess to a complete lack of nostalgia for the two planes. Most of the oldtime pilots became quite sentimental about them. Mike DeBliquy had flown them before the war in

End of Rapide CF-AYE. They didn't fly worth a damn but they sure did burn good.

Russ Baker's Junkers at the Vancouver seaplane ramp, 1948.

The growing fleet.

Sullivan Bay, a floating village owned by Myrtle and Bruce Collison, became our most important mid-coast base.

Europe, and again when he was flying for QuebecAir in Eastern Canada. The Rapide had a long narrow nose, and the pilot sat up there all by himself. He had to turn his head right around and look back to see the rest of the aeroplane. Mike described it as being "like sitting up there in the lap of Jesus, with all those nuts and bolts following along behind!"

Even putting the Grumman Goose into service did not really solve our problem. The Grumman had better performance than the Rapide, and could fly on one engine, so the safety factor was improved. Also the amphibious feature enabled us to land on the runway at Sandspit, which obviated the need to truck passengers down to the seaplane dock at Alliford Bay, saving about an hour turnaround time. But the Grumman was only licensed to carry a pilot and five passengers, according to Canadian DoT Regulations.

At this time we got our first "International" scheduled licence, to fly between Prince Rupert, BC and Ketchikan, Alaska. The licence was shared on a reciprocal basis with Ellis Airlines in Ketchikan, who also operated a Grumman Goose for this run. We arranged to operate on alternate days. This should have been perfect. But—the Federal Aviation Administration (FAA) in Alaska permitted a much larger gross load—a pilot and *seven* passengers. This fact would be driven home to our great embarrassment, when, as frequently happened, a party of seven would book one way on Ellis Airlines, to find only five could come back on the same plane when operated by QCA. The reason for this anomaly was purely scientific, as we discovered when we presented it to our intellectual superiors in Ottawa. The reason, as they explained it, was that the cold air in Alaska was heavier, and aircraft performance was consequently much better. I had occasion to cross over the Alaska border many times by air, and try as I might, I was never able to verify the DoT's contention it was thicker coming down than it was going up. What my experience did prove was that the regulations enabled the US operator to make a profit which we couldn't, and eventually we were obliged to cede the licence to Ellis Airlines.

The air back on our own side may have been thin and puny according to the DoT's version of the world, but we had it largely to ourselves. CPAL occupied the niche above us and gave us few worries, although we greatly coveted their base at Port Hardy. BC Airlines occupied the niche below us and seemed happy working charter in and around our main routes.

BC Airlines had a similarity to QCA in that it was also started by a non-pilot. Bill Sylvester was operating an automotive U-Drive business in Victoria before the war and somehow just got the idea that an airline would be a nice thing to have. So he took a few flying lessons, and in 1942 put some spare money into a tiny little Luscombe two-seater. But here's the funny part. He didn't fly the plane. He just took the wings off and stored it in his U-Drive garage. Then he bought another plane, a Waco, and put it in storage too. In 1944 he collected yet another Waco. But it wasn't until 1945 that he actually got his first charter licence, and it was later still before he started getting the planes signed out ready for service. By this time he had an Air Force veteran named George Williamson to fly for him and had a commercial pilot's licence himself, and he slowly started taking charters. This was how to start an airline the careful way. Sylvester stayed careful. He stuck to charter work and slowly built up his fleet. He collected a good bunch of pilots and hung onto them — Bill Waddington, Denny Denroche, Bill Bullard. I used to envy Bill Sylvester because his life seemed so simple compared to mine.

There was relatively little friction between us. I knew Bill and we got along fine. In fact, at one point when we were otherwise occupied, we got him to come in and take over most of our Class Three and Four charter licences on a temporary basis. The only problem came when we tried to get them back, but that was more the ATB than Bill. They wanted us to get out of the small stuff by that time. Besides BCAL there was Associated Air Taxi, the outfit started in 1946 by my old tormenter Bob Gayer after he'd given up on his mine. Bob was ambitious enough, and a very creative person — the original "idea man." But he had his problems making a go in the territory that was left to him and was always teetering on the brink of wipeout. He managed to pick up Class Two services into Pender Harbour and the Gulf Islands, and one to Gun Lake, but most of his business was charter. Jack Moul and Slim Knight, two Air Force veterans, started Port Alberni Airways in 1946 and struggled along for a few years before they amalgamated with Bob in a share trade. There were charter outfits being started up from time to time by bushpilots or Air Force veterans but they didn't last.

We came up against one of these in the spring of 1948 when a guy named Russ Baker showed up at our maintenance shop on Sea Island with a beat-up Junkers he desperately wanted to get flying again. He was full of big talk about all the important people he'd had in his planes and our guys didn't like the look of him, so they asked me if they should do the job. I said no way.

I knew this Baker. In 1946 he had beat us out of a thirty-thousand-dollar Forest Service contract flying forest fire patrols in the interior of the province. At that time he'd been just starting his own outfit, Central BC Airways, out of Prince George. For planes I think all he had was one of these war-surplus Cessna Cranes we'd bought but weren't allowed to fly. He might have had a couple of them—you could get the used-up ones for a few hundred bucks. His only good asset was his partner, Walter Gilbert. I last saw Walter when we were both invited to a ceremony at the aviation pavilion at Expo 86. He won the prestigious McKee Trophy as one of the early arctic bushpilots and had a lot of experience running flying operations for the old Canadian Airways, and it was largely due to his reputation they got this forest service contract.

We still couldn't see how the Forest Service could accept them, because they didn't have an aeroplane to fly it. Then we discovered Baker was using a beat-up old Norseman on floats owned by the RCAF. He had appealed to the government or the Air Force, saying this is a matter of national interest, to save the forests, so with a special order they let Baker use this military plane. Well, he immediately began using it on all sorts of other commercial work. On June 22, 1946 it was reported to our directors meeting that the mysterious RCAF Norseman, piloted by a character named Pat Carey, had been caught poaching on our Rupert–Stewart business. This was just outrageous because Baker's charter licence didn't permit him to come anywhere near salt water, but they kept this up for weeks. Our mine-haul customers were telling us, "Get lost. There's a guy with a Norseman here who'll fly for half what you want." We reported all this to the DoT, that this guy named Pat Carey was violating our territory using an unregistered military plane, Air Force #347, and meanwhile the fire patrols weren't getting flown. DoT said, "Tut tut tut, we don't believe it." Be darned if we could get anybody in DoT to go up there and do anything, so Baker flew all summer loading up with cash business in a free plane.

We did a lot of flying for the forest service on the coast and kept going after the interior contract, but we couldn't get it away from him. He had the forestry people in the palm of his hand. In 1949 he got the contract expanded to cover Kamloops and used it as a base to steal traffic from Kamloops Air, another little charter outfit run by a veteran named Harry Bray. Baker had his Beaver by this time—the first de Havilland Beaver ever sold, CF-FHB, now in the National Museum of Science and Technology—and he would swoop down in the Kamloops licence area and load up with Bray's customers at half fare. Bray screamed to the DoT, but, far from pulling in his horns, Baker turned around and applied for his own licence in the Kamloops Air Service's territory. He claimed he was always having to come in and do Bray's flying because Bray's little Seabees couldn't handle the work. Bray came to Bud Lando for help and Bud got Baker on the spot at the resulting ATB hearing. He said, "In the time of Queen Elizabeth you would have been knighted for piracy."

There is a story told by Baker's wife Madge that he stood up, swept off an imaginary tricorne, bowed and said, "Thank you, Sire, *all* my ancestors were pirates!" This strikes me as more the kind of thing he would say he did than something he would really do, because Baker was always very shy in front of a crowd. He was very confident and voluble one-on-one but put any kind of a group in front of him and he would get tongue-tied. When it came to hearings he would always have some one else do the talking. Anyway, whatever was said, it didn't impress the ATB. Baker lost the application. But that still didn't help Kamloops Air. He kept right on poaching and eventually put Bray under. I considered myself fairly creative at interpreting regulations, but this kind of behaviour appalled me. I couldn't believe anybody would have the baldfaced gall to try it, and even less could I believe they would get away with it, but I gradually realized Baker was making a career of doing exactly that.

Within a short time we heard Baker had pulled some sort of a doublecross and cut his partner out. It fitted. We just wondered how anyone with Walter Gilbert's brains had got mixed up with a bullshitter like Baker to begin with. Baker wasn't liked or respected by the airport crowd. He was a real publicity hound and you'd be reading all these cooked-up stories in the papers about his exploits. Wild stories about the Headless Valley and everything else. John Condit does a pretty good job cataloguing some of Baker's BS in his history of Pacific Western Airlines, *Wings over the West*. Baker claimed to have discovered a new eleven-thousand-foot mountain range nobody else ever saw before or since. He told Pierre Berton he had ninety-five hundred hours flying time, which he claimed was twice as much as any other bushpilot in Canada. Baker's own partner Walter had almost twenty years more flying experience and we had numerous pilots with more than ten thousand hours, but Berton ate it up and so did his managing editor at the *Vancouver*

Sun, Hal Straight. The *Sun* made it a personal mission to turn Baker into a folk hero. Once Berton reported that Baker had made over two thousand mercy flights. This would mean that every man, woman and child in his territory had to be rescued at least once by Baker, and some twice, since the entire population of the area was only about fifteen hundred. Once he told the papers he'd made a daring rescue of eight starving miners, only to have the mine boss, Emil Bronlund, later say they had called him in by radio and hadn't been out of food at all. In 1948 he got some kind of a cracker-box medal from the US Air Force for supposedly having rescued twenty-four USAF crewmen downed in the Yukon during the war. Again it was written up all over the front page of the *Vancouver Sun*, even though it turned out over a million of these medals had been handed out between 1941 and 1947. The *Sun* was under different management in those days and was a very flamboyant paper. Everyone referred to it as "that yellow rag." Later the veteran bushpilot Sheldon Luck told me it was all bull anyway—it was him who found the USAF crash, not Baker.

But in 1948 Baker was desperate to have our shop patch up his Junkers. He'd had three of these old Junkers he'd got for next to nothing from CPAL, but Pat Carey had cracked one up, they'd run the other two into the ground, they'd cracked up their Cessna Crane and now they didn't have planes to keep up their licence or their contract. Baker had a backer named Karl Springer who'd been pouring money into Central BC Airways for two years and getting nothing back, so he was anxious, too. One way or another they got us thinking we would actually be dealing with Karl Springer if we took this job on. Springer I could go for. He was very well known, he had developed Granduc Mines among other things, and was referred to as a millionaire, which in those days was still a mark of some distinction. So I told our maintenance people to go ahead and overhaul this plane. We did everything Baker wanted, then he stiffed us for the bill while Springer went out and bought him the brand-new Beaver for thirty thousand dollars. That was my introduction to the dynamic duo of Russ Baker and Karl Springer.

We'd hear rumours too, about how they planned to take over and run all the rest of us out of the air, but it seemed more pathetic than anything else. Fortunately, Baker was well out of our territory. He couldn't legally come near us because he didn't have a licence to fly out to the coast and we couldn't see the ATB ever giving him one, so we felt we had the luxury of ignoring him.

Flying Elephants

THIS WAS THE FUN PART OF THE THING, the period between 1946 and 1950, when we were collecting a pretty good team of people and the company was doing nothing but go ahead. It was strange. By 1950 we had 160 people working for us and revenues of .8 million but in some ways my job was easier than when Hep and I were breaking into business and worrying about hiring our first part-time apprentice. The financial reports I looked at all had a lot more decimal places than they used to, but you got accustomed to that. I had to consciously remind myself that less than ten years had passed since I wrote Hep gloating about selling two fifty-dollar Stromberg Carlsons in one day, and since I'd felt it necessary to write my father on the occasion of seeing my first hundred-dollar bill. I began to have a little less pressure on me and felt I could relax and do a few things on my own, like go home on the weekends, and get my own pilot's licence.

We acquired the little Stinson Station Wagon, which became my personal plane more or less, as I flew it around to local airport committee meetings and filled in with the odd radio trip. I enjoyed flying, but I shiver to think of some of the scrapes I got myself into.

I particularly remember one trip in that first Stinson, CF-FYJ. I had to go to Tofino to see our agent, Ian McLorie, and arrange a contract with a local operator to run a regular limousine service between Tofino Airport and the seaplane base in Tofino Harbour. This was when we were flying Anson land planes Vancouver to Tofino and transferring to Norseman floatplanes for distribution up the west coast of the island. It was a nice day and Johnny suggested I take our secretary Audrey Gerrard along for the ride so she could see some of the places she was writing all the ATB reports about.

The flight over took less than an hour and Ian welcomed us to his house for lunch, after which we rode to the airport in the new limousine service. When we got back to Tofino the fog had set in. Typical west coast fog — visibility less than a hundred yards. We hoped it would clear, but no luck. Ian said the fog would lift by 10 A.M. the next day — it always did — so we stayed over.

It was still socked in at noon. I radioed in to say we would have to stay put but Johnny Hatch wouldn't have it. He needed FYJ for shuttle traffic and he wouldn't accept a little dab of west coast fog as an excuse for holding it up. The fog in Tofino, he said, was caused by cold inlet air coming down Bedwell Inlet and meeting warmer sea air, and all I had to do was taxi back up Bedwell Inlet on my compass a little way and I'd be out of it.

Ian wanted to come back to Vancouver with us so I put him in the front to help me navigate and relegated Audrey to the rear. Then we started out, slowly feeling our way. To make things harder, the channel out of Tofino was a narrow one, passing through several miles of sandy shallows and marked on one side by a line of pilings placed about a hundred yards apart. The fog was so thick I would lose sight of one piling before spotting the next, depending solely on the compass to keep on course. I asked Ian if I should keep the pilings on my left or my right. He said keep them on your left. I said okay, but I kept worrying about it. A bit later I asked him if he was sure. Of course, he said, he looked out at these pilings every day of his life. Alright, I said, and we kept slowly chugging along. There was a strong current running against us and I wasn't making very good headway, but that suited me. I didn't want to go any faster. Then Johnny called and wanted to know if I'd gotten off yet — they were waiting. I told him about the current and he said I should use more power and "get her up on the step." It would be safe enough now that I'd got the feel of what I was doing.

With my heart in my mouth I opened the throttle until the plane lifted up on the step of the floats, almost at takeoff speed. Now the pilings were flicking by like railroad ties. I still couldn't see a thing ahead. I looked nervously at Ian, and he nodded reassurance — yes, yes, you're okay, just keep going. Then I did see something ahead. It was a group of seagulls in the water. But there was something about them that bothered me. They weren't swimming. They were standing.

I looked down, and my gosh, I could see the clamshells in the sand. We only had about six inches of water under us and it was getting shallower by the second. I was so close to takeoff speed, I had the impulse to speed up just a little more and fly the hell out of there, but something made me chop the throttle instead. This had a remarkable effect. As soon as the floats sank back down they bit into the sand and the whole plane stood right up on the ends of the floats. It balanced there for one long, long second, the propeller whirling only inches from the water, then flopped back down on the floats again. The seagulls flew off screaming indignantly in all directions.

Fifteen minutes later we were all out walking around on the sand. Of course the tide was falling, and we were stranded there for four hours. And of course, the fog lifted immediately. I looked around and there, directly ahead of us, was a small island with trees about a hundred feet high. If I had given in to the urge to take off, we would have been crow food. While Ian and Audrey amused themselves by playing noughts and crosses in the sand, I gave inward thanks for the blind impulse that moved the throttle down instead of up. Overhead, our Ansons streamed by all afternoon on their approach to the airport, gleefully dipping their wings at the boss on his sandbar down below.

Every day there would be *something*. Once the phone rang and the receptionist answered it, frowned and carefully said, "Yes, ma'am, we can fly to Vernon." Bill Peters pricked his ears up, I put down whatever I was doing, the receptionist frowned a bit more and said, "Just a minute please," then turned to us. "A lady wishes to know if we can fly a musical adjudicator to Vernon tomorrow morning."

"What in hell is a musical adjudicator?" I whispered. Everybody in the office shrugged.

So Bill hollered across the room, "Get the weight and the height and the width and we'll see if it'll go through the door of a Norseman!"

We had in mind a jukebox or something. It turned out to be a delicate white-haired lady from the New York Philharmonic who was going up to inspect the talents of Vernon music students.

Art Barran came in one time with a weird tale after flying a Strannie into a remote lake behind Prince Rupert with a trapper and his outfit, which included a full team of sled dogs. We took all the seats out and loaded all his feed and his traps and his dogs in there

My Stinson high and dry at Tofino. Note island in background.

Slightly unorthodox landing by John Hatch at Garibaldi left passengers unperturbed.

My Stinson stuck after snow-float landing on Garibaldi Lake.

Norseman CF-CRS stuck on Garibaldi Lake. Al Alsgard applies crisco to the plank toboggan as Johnny Hatch looks on.

and everything went fine until they tried to land. A flying boat makes a heck of a racket when you touch down because you're right in the thing, and there's no soundproofing. When there's a little ripple on the water it sounds like you've landed on a corrugated iron roof. As soon as the Strannie touched the water all the dogs flew at each other's throats. The whole cabin area was just a whirling ball of fur, teeth and blood.

Another time we received a charter for Garibaldi Park. Three young lady hikers wanted to be taken into Garibaldi Lake with their packs and supplies. They would go up to the meadows and stay in a log cabin the Parks Board kept there for mountaineers. The weather report was marginal but Johnny Hatch decided he would try making the flight himself. All went well until he turned out of the valley and headed up toward the end of the lake. The end of the valley is blocked by an eight-hundred-foot-high lava "dyke" that forms the lake, and he had to fly up over this barrier to land on the water. Normally this would be no problem but on this day there was a solid deck of cloud hanging only about two hundred feet above the lake, leaving only a narrow slot for the Norseman to squeeze through.

Just as Johnny was approaching the barrier, a violent downdraft of cold glacial air hit the aircraft and caused it to lose several hundred feet of altitude. It was impossible for the Norseman to gain enough altitude to clear the barrier in the short distance left. It was also too hazardous to attempt a tight turn at low speed in descending air currents, so Johnny took the only course left to him. He chopped the throttle, shut off the ignition, and aimed straight for two medium-sized, springy-looking fir trees. The aircraft, now at stall speed, struck the trees at about forty-five feet above the ground, pushed them over to about a forty-five-degree angle, then slid down the trunks like an elevator and made a reasonably soft landing.

Quite a few things happened to the aircraft during the process. Both wings sheared off. The pontoons doubled back under the fuselage like pretzels and the engine came off its mount. Gasoline from the wing tanks was everywhere, but fortunately nothing ignited it. When the broken branches and glass and bits of aircraft stopped falling, there was complete silence, and Johnny looked around to see how his passengers made out. Before he could think what to say, one of them turned from the window where she had been studying all the moss and rocks and trees and exclaimed, "Oh, isn't this absolutely *bee-yootiful*!" None of them had ever been in an aircraft before and they had nothing with which to compare this uncommon performance. They seemed to assume that this was just the normal way you landed your floatplane on a mountain.

Johnny apologized for the inconvenience and explained they would have to walk the rest of their way to the cabin, but he would help them with their stuff. He didn't have anything else to do at the moment. As luck would have it the aircraft had come down on the main Parks Board trail, which led from the Parks Board cabin down Rubble Creek to Squamish. It took about fifteen minutes to reach the cabin. Johnny told them that when they got back to Vancouver they could apply at our office for refunds on their tickets, but as far as I know they never did.

The Parks Board we did hear from, however. They accused us of wantonly blocking off their main trail, scuffing up two prime second-growth fir trees and causing an unsightly mess. We flew a crew in, landed normally on the lake, walked down the trail, cut the aeroplane into small pieces and heaved it all over the edge into the canyon where we hoped nobody would find it for a long time. Nobody has yet, that I've been told.

Another time it was very dry. There hadn't been a spot of rain for weeks, the woods were shut down, everybody's lawn was brown and the big question was, how much longer would it go on? Some bright egg around the office got the idea we should go into the rainmaking business. As a publicity stunt. We looked into it and found out what they did was fly above the right sort of cloud at just the right time and seed it with granulated dry ice. Fine. We contacted all the papers and radio stations, told them what we planned, they liked it and the deal was on. We just had to wait for the right sort of cloud. Well, we sat around staring up at the sky until we were dizzy, and couldn't find any sort of cloud at all. Then finally a wispy little thing that looked like you couldn't squeeze a drop out of it appeared over Burnaby and we decided to give it a try. The media appeared on cue, I posed holding this fuming container of powdered dry ice in front of one of the planes, where the cameras couldn't miss the QCA insignia, and we took off. Bill Peters was the pilot. By the time he manoeuvred into place over this low bit of cloud the ice powder had welded itself together in the bucket and wouldn't sprinkle delicately out through the camera hatch like we'd planned. I tipped it up and tapped the base like you would a ketchup bottle, but nothing came out. By this time we'd overflown this miserable little slip of mist and Bill told me to hold off while he made another pass at it.

"Okay!" he said, as we came back over.

I tipped the bucket out the window again, but this time I wound up and delivered the bottom a real whump. This succeeded in loosening all of the ice at once and it dropped out in a solid slug. The last I saw of it, it had sliced through the cloud and was smoking toward the rooftops of Burnaby like a cannonball. We landed back at the airport not knowing if we'd be cheered for making rain or arrested for mass murder. As it happened, the reporters had all retired to the press club and nobody at the office had received any reports of Burnaby roofs being caved in, so we slunk back to our desks with nobody the wiser. About two nights later there was a light rain and we received several very nice writeups giving us full credit for our public-spirited work.

We had never stopped to think about publicity much. It was generally one commodity we had lots of, both the good variety and the other. But now that we were between crises we had time to think about some of the fine points of being in business, and getting publicity was one of the things that came up. Norm Landahl, who started out in 1946 as "an accountant who isn't afraid of getting dirt under his fingernails" and had now become something like office manager, generally took care of any publicity concerns that came up. I was often called upon to rush over to the loading ramp and shake hands with some celebrity or other who was taking a ride with us. I remember one time they came after me in a big sweat, made me change into a fresh suit, sped me over to the tarmac to meet a plane and it was Gracie Fields. She seemed nice enough, we had a chat and off she went. Then reporters came up to me wanting to know everything about her and I had to admit I didn't even know what she did. I was only vaguely aware she was some kind of a name in show biz.

As time wore on and getting the right sort of publicity became more of a concern, we hired an agency, O'Brien Advertising, to take some of the haphazardness out of it. But haphazardness had its own way of creeping into everything we did. I remember one time in particular. The year was probably 1949, but it's not important. What was important was that the famed Ringling Circus was in town. At QCA, in the meantime, everything was going along humdrum and normal. So humdrum that Johnny Hatch and Bud Lando were saying it was about time we were doing something to get a little public attention. We could run large ads, of course, but this cost money—far more than we had to spend. What we needed was *free* publicity. We had empty seats to fill. This was all very much on my mind when this guy came out. He gave his name as Ray Munroe, just as though I should instantly recognize him. The name meant nothing to me, so he said he was in public relations and was currently handling the advertising and promotion for the Ringling Circus, and he had a plan that would not only be good for the circus, but would spread the name of Queen Charlotte Airlines from coast to coast. In fact, he said he would guarantee it would make the first page of LIFE magazine. This got my attention, so I said, "Go on, what do you want me to do?"

First, he wanted to know what was the biggest aeroplane we had that could fly to Victoria? I told him the Stranraer, of course. What was the heaviest thing it could fly? I said up to two tons or more, depending on size and shape, what did he have in mind? He swore me to absolute secrecy, as he explained that if this got into the newspapers before tomorrow we were dead. He took a grip on the arms of his chair, leaned forward and looked right in my face and said, "We want to fly an elephant from Victoria to Vancouver tomorrow. Can you do it?" He made it very clear from the start that we were in it solely for the publicity, so we didn't talk price, even when it became apparent that we would possibly lose a full revenue trip because of it. I asked him for more details—maximum height, width and weight of this elephant, and he said it was only a baby—no problem—but he would ask the elephant's trainer to come in and I could get all the gen from him.

The elephant trainer came in. He was only a little guy, not even my size. Didn't look big enough to handle an elephant, but I guessed they must know their business by this time so I fired the questions at him and by golly, he had an answer for every one. I couldn't corner him. It went something like this.

Q. How much does he weigh?

A. About twelve hundred pounds.

Q. What is his height?

A. About five feet, six inches.

I said no way, the door is only four foot.

He said that's okay, one of his tricks is he can walk on his knees. He does that so kids can climb up on him.

Q. H'mm. How wide is he?

A. About thirty-six inches across the belly.

I said that does it, our door is only twenty-six inches.

He said no problem at all, another of his tricks is he can suck his stomach in when I tell him so we can put his harness on.

He had an answer for everything. I could just see this elephant going up our loading ramp into the aeroplane in sight of all our regular passengers, on his hands and knees with his stomach sucked in. I shook my head. Munroe said, "Look. We've a hell of a lot to do to get ready before tomorrow morning. Let's get rolling. Now here's the scenario: we will go over to Victoria on the night boat with the elephant, and tomorrow morning we are introducing him to the premier of British Columbia, W.A.C. Bennett. He likes elephants you see, and this will be broadcast on all the networks, radio and television. Then while the Premier is shaking hands with the elephant the trainer gives a signal and the elephant falls over on his side in front of the premier and makes groaning noises like he's in pain. Well, now everyone rushes around and calls a doctor. We have one standing by, of course, and he comes rushing in with a stethoscope and does an examination right there on television. Then he gets up off his knees and announces that the elephant has acute appendicitis and they have to get him to a hospital right quick. But the vet says there is no animal hospital in Victoria that handles elephants. The closest one is in Vancouver, but he doesn't think the elephant will live long enough to survive the steamer trip. Can they arrange to fly him over? Someone pipes up and says to call Queen Charlotte Airlines. They are the only people that could do it—so someone rushes off to the phone to call QCA and comes back shortly to say there will be a plane right over, and everybody including the premier is impressed all to hell." In the meantime, we have one of our aircraft all prepared and in Victoria Harbour set to go. Munroe says he doesn't care how much it costs us, the publicity we will get will pay it back a thousandfold.

And so it was set up. I would have a crew work all night on one of the Stranraers, and we would have to cancel or delay the trip it would normally be taking by four hours. The traffic department would simply have to cope somehow. Then I called Charlie Banting and filled him in on the plan. To say the least, he didn't think much of it. He said, "Christ, that thing will go right through the floorboards and probably right on through the skin of the hull. She isn't stressed for that kind of a load!" I said, "But Charlie, it only weighs twelve hundred pounds. That's only three hundred pounds on each foot, and some *people* weigh that much!" Charlie said, "Fine, but what if he decides to stand on one leg while he scratches himself or something?" Without the trainer around, I felt that Charlie had a valid point, so it was decided to beef up the floor. This meant removing all passenger seats in the centre compartment, ripping the existing floor panels up and building a framework of wood. For this we used some three-by-twelve timbers we had intended for the cabin at Garibaldi. Then we installed securing points on both sides of the hull and made up some heavy canvas webbing to put around this elephant like seatbelts and shoulder harness. Then, just to make sure, we got a bale of hay and spread that on the floor. By about 6 A.M. we were ready. Charlie had one shift work all night doing this, and I congratulated them on the terrific job they had done on short notice. When I explained to them the purpose of the undertaking, they sort of looked at each other as though to say, "What's he been sniffing?" They wouldn't believe the elephant bit. I said wait and see. I had tried to get hold of Bud Lando and Johnny Hatch so they could share the experience, but both had gone out, and time was so short I couldn't get them.

It was arranged that Ray Munroe was to call me by 8 A.M. to settle on the time for the flight and take care of last minute details. No word from him. I was getting anxious so by 9 A.M. I started phoning his office. "I'm sorry, Mr. Munroe has not come into the office yet, can I give him a message?" I said for them to get hold of him wherever he was, and whatever he was doing, this was very urgent, and extremely important. But no one heard of the SOB again. Johnny came raving in and said, "What in hell is going on around here? Only one Strannie flying and the other all trussed up like a battleship, and Charlie's so mad he's spitting nails!"

For the next month the boys in the hangar tried to avoid looking at me, as though I had completely flipped my lid. They never did believe the elephant story. I eventually gave up trying. To top it off, our advertising counsel, Jack Markay of O'Brien Advertising, dropped around on his regular call and I told him what had happened. He went straight to the telephone and kept it jumping for about half an hour before he got an answer. I guess he knew numbers to call that I didn't. Anyway, he said he finally ran the story to earth. Munroe had had this brainwave and sold himself on it, and also me, of course, but his clients decided at the last minute to scrub the deal. Munroe had proceeded to get very drunk, and he was still drunk. Then I got a real lecture from Jack Markay. He said, "What do you think you're paying me for? Why didn't you call me right at the start? I know that bastard from way back, and I'd never have let you get into this mess."

During this time I was very involved in trying to get an airstrip built at Powell River. We were flying into Powell River but we were landing on Powell Lake with Stranraer flying boats and Norseman seaplanes, and we wanted to get on land. We had the Board of Trade, we had the municipality, we had the local flying club, we'd meet once a month in the old Westview Hotel to try and promote this thing. This particular day I had to be up there at twelve o'clock so I flew up in my Stinson and it was a fine day, a beautiful day, you could see to the ends of the earth. After the meeting I found myself talking to Al Alsgard, publisher and owner of the *Powell River News*.

I had to fly on up the coast this day and do some repairs at the Willcock Store on Stuart Island so I said to Al—I'd known him since we were kids together on Savary Island—I said, "Al, I'm going up all by myself, why don't you come along for a ride?"

"My gosh yes," he says. "Fine. Do I need anything?"

"No, nothing. Come just as you are. Bring your camera."

We drove down to the Lake, got in FFW, took off, went up to Stuart Island, flew back and my god, the view was overpowering. I got up to a fair altitude and you could see clear over the Gatineau Range right back to Garibaldi. It was the sort of day that just put mundane cares out of your mind and fired your imagination. I had an inspiration. I turned to Al and said, "You know, Al, what I really feel like doing right now is flying straight over there to Lake Garibaldi. How would you like to come over with me, then come into Vancouver and I can put you on Flight Three in the morning. You'll be back in your office by nine A.M. What do you say?"

"Well, how'll I let the wife know? She'll be worried," he said.

"Oh heck, that's no trouble," I said. So I got on the radio and called our agent in Powell River, please advise Mrs. Alsgard that Al Alsgard will be in Vancouver overnight, returning tomorrow morning. This was checked okay and Al was happy.

Now, I wasn't being entirely frivolous. I had a reason for wanting to make this trip. We had a forty-year agreement with the Garibaldi Parks Board to develop air transportation between Vancouver and Garibaldi Lake, and we had built a lodge in there. We flew in with seaplanes during the summer, but it's a high, glacial lake, ice-free only a comparatively short time, and the question we were hung up on at this point was how to get in after freeze-up. You might think we could use skis and land on the lake, but to land on skis you have to take off on snow, and we were taking off on water.

So how do you take off from water and land on snow? This was the problem we were mulling over in our idle moments.

We might have gone on mulling for years, but as they say, necessity is the mother of invention, and the previous winter we'd found ourselves in a position where we had to find an answer. This was one of those very harsh winters we seemed to get a lot of in the late forties and early fifties, when the temperature at Vancouver Airport stayed around zero for weeks and the Fraser River froze over. We woke up one morning and found ourselves out of business because our seaplanes couldn't take off on the ice. They were all beached on dry land and we realized that if we could get them over to our base in Nanaimo Harbour we could shuttle passengers back and forth between Vancouver and the Nanaimo airstrip using a wheel plane and continue flying up the coast with floats, but meanwhile, how do we get our seaplanes out of Vancouver Airport?

It was Johnny Hatch who hit on it. There was a good cover of snow alongside the runways and he taxied a floatplane out of the hangar, pulled off the beaching gear and sat the bare floats down in the snow. He taxied around, it moved along nicely, then he pushed the throttle forward and took off. Nothing to it. So we took all our seaplanes off in snow and by golly we landed them on snow at the Nanaimo Airstrip. We didn't even have to use Nanaimo Harbour; we just met the land shuttle at the airstrip and flew up the coast directly. It worked fine. Except of course it was illegal, and we didn't dare tell the Department of Transport about it.

It wasn't long before someone said, "Here's the answer for Garibaldi Lake! We can just land in the snow on floats! Fly in there all year round." The only question was how the snow was going to be. Would it be level enough, or all hummocky? Would it be ice or powder? This was burning a hole in my head. I wanted to get on with the thing. It was now June and we only had a few weeks before the lake ice started to break up for that year.

I explained all this to Al Alsgard and said, "This is what I want to find out today. I just want to take a low pass over the lake and see." He was very interested. He got his camera ready. We went in over the top at ten thousand feet; you could see all the way to the Rocky Mountains. I found Garibaldi Lake, circled in past the Sphinx Glacier, did a beautiful let-down over the Sentinel Glacier, leveled out and flew low over the surface of the lake. It was as smooth as a billiard table, sparkling, unblemished white. I pulled out at the end and went up.

"Al, are you nervous?" I said. "Because I'd like to put my floats down on that surface, just to feel it. I want to know how solid it is."

"Go ahead," he says. "I'm having a wonderful time."

I went around, came down over the glacier again, did a let-down, a bit of a sideslip and levelled off about twenty feet over the lake. Then gradually, gradually let down till the floats were skimming the snow. It was so smooth I couldn't tell my floats were down. I eased the throttle back. It was soft, soft powdery dry snow and we glided to a gentle, perfect stop. I shut the engine off. It was stunning. The sun was brilliant, the snow was sparkling, the lake was impeccable. We opened the window and outside of the engine cooling making a ticking sound and gyro instruments still spinning it was the quietest thing you ever heard in your life!

"By God!" Al said. "What a country! I'm going to get a picture."

He was just in his light clothes, no overcoat, and Al was a big man. When he stepped off the float with his press camera, he went in up to his armpits.

Just powder and air.

But he got his picture, came back in and said "Boy, that's enough for me." It was sunny but it was still damn cold. I'd found out everything I needed to know so I started up, opened the throttle gradually—and nothing moved. I opened it full up. Still didn't move. I rocked it. I tried the ailerons. I tried everything, but the aeroplane wouldn't budge. I sat and thought about it for a minute. Somewhere, sometime I'd heard bushpilots talk about skis warming from the friction of landing, then freezing into the snow. I thought I remembered what you were supposed to do. I turned to Al.

"I hate to ask you to do this, but you will have to get out again and take the aircraft by the tail and shake it up and down while I open the throttle. If she starts moving, try to scramble up on the float. If you can't catch up I'll circle back." He went out, I opened the throttle, he shook, and she started to move. And he couldn't get near it. He was left there looking like a lost snowman. It blew powdered snow in his pantlegs and out his collar. Oh, he looked miserable. I went around in a big circle half a mile wide, came back and missed him. I went as slow as I could without getting stuck again. It took me about three tries. Finally I got slow enough and close enough he was able to lunge and catch a strut and pull himself up onto the float while I put on power, and he clambered into the plane.

"B-b-boy, will I ever be glad to get w-w-warm," he said. I opened 'er up and waited, and waited, and waited—and eight miles an hour was the best I could do.

I kept going and headed for the lodge. The lodge was a

three-storey cabin and the peak of the roof and one little window was all that was showing. There was about twenty-eight feet of snow. I stopped the plane, wallowed over to the cabin, got the window open, got the oil stove going and got poor Al Alsgard in there. I had further plans for him, but I wanted to get him warm before I told him. It was pitch dark inside, so I lit a lantern.

"Al," I said, "it may be that I can get airborne with just myself aboard. I'm sure if I can get back to base and talk this over with the boys we'll be able to come up with something, perhaps a different aeroplane with more power. But you're going to have to shake me loose again."

Well, what choice did the poor bugger have? I got him out there while we still had some light, he waggled my tail, I got moving, and bygosh, without Alsgard's extra weight, I got her up to speed and lifted off.

I'd been in touch with Vancouver all through the ordeal, keeping them posted, and they had been taking it all without making any comment. When I taxied up to the float in Vancouver there was Johnny Hatch standing there, looking very grim. He had originated the floats-on-snow technique and he had wanted the honour of testing it on Garibaldi Lake. He didn't say anything, but if he had said what he was thinking I suspect it would have been, "You're lucky to be alive, you damn fool," or something along those lines. He was very brittle with me.

Johnny had our best-performing Norseman CF-CRS ready to take off, and didn't want to hear any of my explanations. He had it all figured out. "The difference between your aeroplane and CRS is you're not supercharged," he told me. "That lake is five thousand feet up and your darn little Lycoming motor only has half its power at that altitude. There'll be no trouble in CRS because it's supercharged and it's got full power at five thousand feet." I was still carrying my briefcase, still dressed for my meeting in Powell River, but there was no time for anything but to climb in and take off to rescue Alsgard.

It's a twenty-five-minute flight from Vancouver straight over the Lions to Garibaldi Lake. Johnny said, "Now watch. I'm just going to put the keels of my floats in and feel it." I couldn't bring myself to tell him that was exactly what I'd done. He comes down over the Sentinel Glacier, levels off, touches the keels of his floats, which on the Norseman are very deep-veed, slows down, then says, "Good enough, way we go," and opens the throttle up again. And the aeroplane goes slower and slower and slower and stops! The floats are sunken clean out of sight! The Norseman was a much heavier machine than the Stinson and didn't work half as well.

I was vindicated, but much too alarmed to take any pleasure in the fact.

This is a good half-mile from the lodge and we're sitting there trying to appreciate all the implications of these last developments when we notice something moving in the snow away over on the far shore of the lake. It's Al Alsgard, struggling out to meet us, holding his press camera over his head. All we can do is wait. Oh, it took him ages, and when he finally got to us he was dead beat.

"Boy, am I ever glad to see you guys," he gasps. We said, well, we didn't know why because we weren't going anyplace. We gave him a chance to catch his breath while we radioed Vancouver that we were Garibaldi RON — Remaining Over Night — then all three of us started out for the cabin. We didn't know what on earth we were going to do, but we knew we weren't going to do it in what was left of that day. There was something funny with Alsgard and as soon as I smelled liquor I knew what it must be, although I found it hard to believe. There was only one bit of liquor in the cabin, an emergency bottle of rye which I had hidden in the rafters down behind the wall plate and out of sight the previous year. But sure enough, he had gone over that building till he found it, and he'd killed the thing! He was higher than a kite!

We got the fire going again, relit the lamp, and started to look for something to eat. The cabin was well stocked, but of course everything was frozen to the consistency of young granite. There was frozen soup, rocklike pork and beans, concretized canned milk which we thawed for coffee — but once canned milk has been frozen it's very strange, it becomes granular. There were eggs which bounced like golfballs. We didn't feast, but we didn't starve.

Tobacco was another matter. We were all smokers — Al and Johnny smoked cigarettes and I a pipe — and we were through what we had with us before we realized what we were in for. They were soon retracing their steps, collecting butts and emptying out the ashtrays in the aeroplane, salvaging the unburned tobacco and rolling it in writing paper. This didn't appeal to me so I tried filling my pipe with tea leaves dried on top of the stove. I was happy enough with it, but Al and Johnny objected most piteously, claiming the smell of it made them sick.

All night we talked about what to do in the morning and it was obvious to both Johnny Hatch and me that we had to somehow increase the surface area of the aeroplane's floats. We had to construct some kind of toboggans and then contrive to get them under this two-ton Norseman sitting down in this deep powdery snow. We were up early. In the cabin I found a double-bitted axe, a carpenter's hammer, a handsaw, a keg of spikes and a few common nails. There was no lumber, but the spikes reminded me that we had just recently built a dock down at the lakeshore using rough cedar two-by-twelve planks, eight feet long, which we'd flown in. After some exploring we figured out where the dock was and dug a great long shaft twenty feet down through the snow, which was in layers like a layer cake—layers of slush, layers of powder and layers of re-frozen slush like ice. We exposed the boards, pried them loose with a peavey and passed up about ten of these heavy planks. That took the morning. I sawed two of these planks in half, giving me four pieces four feet long, and these I attached at an angle to four full-length pieces so I had two sleds, two feet wide and twelve feet long, with the forward four feet angled upward. I put sides on them for strength and then took some short pieces on edge and veed them out with the axe till they fitted the bottom of the floats like lifeboat chocks. To lift the aeroplane we had to cut trees down for pries and drag out a bunch more planks to act as the fulcrum.

You had to get surface area, so it didn't punch into the snow, then place blocks of wood on that. We had to do this not once but many times, because every morning we'd find the aeroplane had sunk down from the warmth of the sun and frozen in again. It was brutal work. It took us five days, which included one day of blizzard when the thermometer went down to twenty-five below and we couldn't get outside.

Alsgard had started off in a great flap about getting back to put out his newspaper, but this concern was soon supplanted in his mind by one simply of getting out alive. He tried to be of help but he was not terribly practical. Johnny and I did most of the work and Alsgard, I think, prayed. At least he mumbled a lot.

The morning of the day came when we finally had these toboggans finished. We took aluminum sheeting off the cabin roof, which we nailed to the bottoms of the rough planks, and coated this with rancid Crisco. We lowered the plane onto them, it held up, and we decided to try it. Johnny had very thoughtfully drained all the lubricating oil out of the crankcase when we arrived, which we now heated on the stove and carried very carefully out in buckets and poured into the frozen motor. We got Alsgard aboard. Johnny got the engine going and I stayed out to wiggle the tail. Our labours were now about to be put to the test.

These toboggans must have weighed about half as much as the aeroplane. They were the most ungainly-looking things you ever saw. But they moved. I leapt for my life. Johnny eased the throttle ahead, the motor laboured and we very sluggishly gathered speed. We had three-and-a-half miles of lake ahead of us. Johnny had the throttle wide open now. We were doing twenty miles per hour, less than a third of what we needed. All three of us had our eyes rivetted to the airspeed indicator. Twenty-five, thirty, thirty-five—it flickered and faltered, and our heartbeat flickered and faltered, but it started rising again and after an eternity it reached seventy-five. Then and only then did Johnny pull back on the control column and without a hitch the plane lifted into the air. As he circled back we could see the toboggans, still clipping along about thirty miles an hour for the other end of the lake. They floated around the lake for years and I used to see kids paddling around on them having a great time, blissfully unaware of their bizarre history.

We got back to Vancouver, loaded Al with my sincerest apologies back on a plane to Powell River and I went back to my office. On my desk was a telegram from Dan McLean, the director general of civil aviation in Ottawa. It contained a very curt message. The DGA wanted to know if I could show any reason why John Hatch, operations manager, should not lose his pilot's licence, and Queen Charlotte Airlines its operating certificate, for illegally flying passengers in a non-approved aircraft.

This was serious. We were the third-largest airline in Canada, and here we were being threatened with dissolution. What happened was, at this time we were flying in fierce competition with other operators on the coast for the lucrative freight and passenger business into the Alcan smelter project at Kitimat. They used to monitor our radio channels so that when they heard we had passengers waiting at a certain location, they could try to get an aeroplane in ahead of us. They had heard all of our calls from Garibaldi Lake, and seen it as a chance to make trouble for us, which was standard practice. They had reported us to the DoT and made sure it got the attention of Ottawa.

I climbed on a plane and was in Ottawa the next morning. I went in through the green baize-covered door in the Number Three Temporary Building and found Dan McLean waiting for me. The

Overlord of all civil aviation in Canada, from whom there was no appeal. The Godhead. He told his assistant to leave the office and close the door. He was in a serious mood.

"Alright, Spilsbury," he said. "What is going on?"

I played innocent. I said I really didn't know what he was referring to.

"You do know what I'm referring to," he said. "This flying in and out of Garibaldi Lake without properly equipped aircraft."

"We landed in Garibaldi Lake in two different aeroplanes," I told him. "Our Stinson CF-FFW and our Norseman CF-CRS. We landed in both cases on regulation floats and we took off on regulation floats."

"They tell me you used skis," he said, "and that they were non-approved skis."

I said, "Well, I'm sorry but your information is incorrect. We taxied on two toboggans but we never flew with them. I'm not aware that there was any contravention of flight rules."

"Well read your Air Regulations, Spilsbury," he said.

"I have done, sir," I said.

He reached over and got out the little red book, opened it up at page whatever-it-was, and I think I can remember the wording: "Seaplane, definition: A seaplane is an aircraft designed for landing and taking off a water surface."

"Now Spilsbury," he said, "do you see anything there that would give you reason to believe it is permissible to land a seaplane on snow?"

"I see nothing that rules it out," I said.

"How do you figure that?" he asked.

"Well, snow is water, only at a low temperature," I said. "And it doesn't say here what temperature the water is required to be."

Fortunately, Dan McLean had a sense of humour. He tried not to smile and said to me, "This is just a play on words. Now let's get down to it. The fact is, you people have been fooling around with irregular equipment, and what's worse, you've been doing it with a passenger on board, name, I believe, Mr. Al Alsgard. Now, I'm not going to put up with any more of this. If you people want to experiment, apply to us and get an X-licence. Then instead of CF-CRS it's CX-CRS, it's an experimental aircraft and you can risk your neck all you want as long as you don't carry any passengers."

He gave me a good talking-to, then he leaned back in his chair and smiled a slow, fatherly smile.

"You know," he said, "the trouble with you west coast people is you always think you're the first people ever to do anything. It never occurs to you others have run into the same problems and already worked out perfectly good solutions. There is such a thing, you know, as the 'Federal Wheel Ski.' It is an aluminum ski with slots for your wheels to go down and you have a hydraulic hand pump. You pump the ski up four inches so you can land on wheels; you pump it down and you can land on a snow surface. They are approved and with them you can take off from Vancouver Airport and land on Garibaldi Lake and carry passengers to your heart's content. Now for godsakes, quit fooling around and get yourself a set."

So for sixty-seven hundred dollars I bought a pair of Federal Wheel Skis custom-fitted for a Norseman aircraft. We put them on the next year when the lake was again frozen over and Johnny Hatch flew in with our chief of maintenance Dick Lake, a commercial photographer by name of Harold Vandervoort and two other passengers. There were five of them. They circled past the Sphinx Glacier, let down over the Sentinel Glacier, touched down on the lake and plunged down, not just to the depth of the floats as we had done, but right up to the belly of the aircraft. The propeller was in the snow. They were twice as far from the lodge.

Some second guessing had been done by now, and this time it occurred to me to have toboggans built in our shop in Vancouver which could be flown out to the lake. They were constructed out of aircraft-grade birch plywood by our chief engineer Charlie Banting, who did the job in jig time with great flair and precision. We took them out in an Anson along with a lot of winter gear—warm clothing, mukluks, food, even snowshoes—and dropped them from the air. Johnny and his party went to work with their pries and finally, five days later, they got out. But there was a hitch.

Charlie Banting's professionally-built toboggans had been designed to hook onto the tips of the skis with the idea that after they got airborne the pilot could shake them off by doing a quick flip. Well, Johnny flipped half the way home and one dropped off, but the other just wouldn't let go. So how do you land, with a big sheet of plywood over one wheel?

There was some snow on the ground at Vancouver and they didn't know whether to land all in snow, or partly in snow and partly on the runway. Johnny chose the latter, thinking this would equalize the drag better, and landed with the toboggan on the snow and the ski on the grass. He went right up over a snowdrift left by the snowplough, broke the belly out of the aircraft, headed straight into the hangar and slammed the door before the DoT had a chance to see what happened.

The lesson we learned from our experience in Garibaldi Lake was that floats were quite as practical as skis for landing on snow, but it was not practical to let Dan McLean find out about it. We later hauled tens of thousands of gallons of diesel fuel from Stewart in northern BC, taking off in salt water and landing on the glacier at Granduc Mine. We even carried the fuel by pumping it into the centre compartment of the floats. We'd land uphill, stop, pump out the fuel, swing the aeroplane around by hand, and take off down the glacier. We kept the mine going all one winter. This snow-float technique was picked up in Alaska and as far as I know they're still using it up there. But it's still very illegal in Canada.

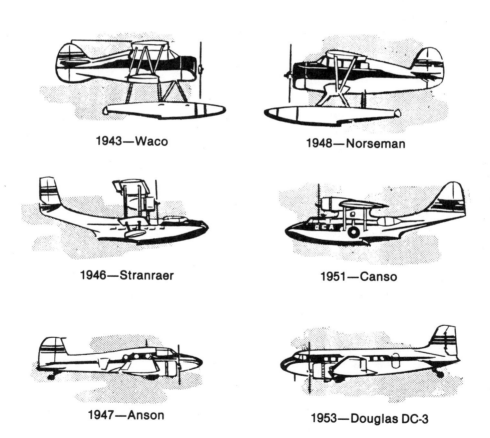

1943—Waco

1948—Norseman

1946—Stranraer

1951—Canso

1947—Anson

1953—Douglas DC-3

QCA: the Queer Collection of Aircraft

NINE

Queer Collection
of Aircraft

FINDING NEW AIRCRAFT to put in service was a constant preoccupation as we grew. Between 1946 and 1951 we were constantly needing to add capacity but we never had much in the way of capital to purchase it with. You couldn't get money from the bank. They wouldn't take a mortgage on a plane. So we had to raise money for new equipment out of operating revenue, or from investors. I had nothing more to invest, now that I'd cashed in my five-hundred-dollar life insurance policy. Nor had Hep or Tindall. Bud Lando, on the other hand, had considerable money within his family. I kept inviting him to buy in, and he did, but only in a small way. When it came to new equipment we could still only look at bargains.

We became so expert in the purchase of used aircraft we set up a separate company, Western Aircraft Sales and Service Ltd., to buy and sell planes as a sideline. Through it we got several dealerships, including one for the new de Havilland Beaver. We would have dearly loved to have a Beaver or two ourselves, but they cost thirty-two thousand dollars. We brought some in for other buyers, including Pacific Mills, the Gibson brothers and Frank Beban Logging, but we could never afford to buy our own. For ourselves we had to make do with things like the two Dragon Rapides for sale in Winnipeg. These were the same planes Johnny Hatch had broken in on in the thirties, when they were owned by British North American Airways. They went broke and Wings Limited bought the planes. Central Northern ended up with them and when they had them worn out they were trying to sell two for the price of one sort of thing. Johnny went back and scouted them and he told me, well if we got 'em they would total about half the price of a Norseman and do, let's say, between them, maybe three-quarters of the work. Deal. I think we got them for two thousand down and five hundred a month.

We were on the lookout for anything we could afford, and we ended up with Stinsons, Cessna 180s, a Bellanca Pacemaker, a Fairchild Husky and a Stagger-Wing Beech, but our mainstays were Norsemen and Ansons. We once bought sixteen Ansons at one go, from War Assets at Abbotsford. These planes, Avro Anson Vs, were designed in England before the war as a light bomber, but when they were succeeded by a superior aeroplane they were converted to training command. Twin engine, plywood skin except for a fabric strip on top of the wing tanks, pack nine people.

I remember when we bought that big bunch. We'd go over to the Abbotsford Airport and we'd drain the water out of the engine, get some gas in, start 'em up if we could, we'd fly them over to Vancouver Airport and work on them. I'd take pilots over there in the club Stinson and I felt really important, but they hated riding with me. I remember Doug McQueen yelling at me one time when I was letting down for Abbotsford Airport a tiny bit fast, "Too much airspeed! Too much airspeed! Pull 'er back, pull 'er back!" They didn't trust me. Another time we went over there and flew this one back with Rupert. I was sitting in the co-pilot seat thinking how wonderful it was to have a twin-engine aeroplane, they're so safe and everything. Then I looked out the window and the fabric on top of the wing was up in a great big blister and POOF! it blew apart. I pointed this out to Rupert with some alarm and he said, "Yeah, that'll all come off."

Some of them we just stripped the engines out of. They had Wasp Juniors, which was the engine in a Beaver or a Goose and very popular. We paid $650 for some of them, complete planes with two engines, and today for a Wasp Junior motor I bet you're looking at twenty thousand. Others we converted into seven-passenger commercial planes for our land runs — Vancouver-Nanaimo, Vancouver-Comox, Vancouver-Tofino. We lined them with fabric and put mahogany panelling on the doors and made them into quite a cute little airliner — we were quite proud of them. We sold two to Baker and some to Tommy Fox's outfit, Associated Airways, in Edmonton and much later we sold six to Ecuador.

Of all the queer collection of aircraft we assembled in the course of our bargain-hunting, it is clear to me the one we'll be most remembered for is the gangling Stranraer, the "whistling shithouses" we set beating their ungainly way up and down the fog-shrouded cliffs of the BC coast, loaded to the gunwales with Chinese second cooks and Finnish chokermen chewing snoose. This is the image that comes to my mind most strongly when I recall the company's salad days, and I am happy to have it as our epigraph.

In a way this is very ironic, because as time when went on we were all rather embarrassed about the Strannies and were never so proud of ourselves as the day we finally graduated to DC-3s and left the old flying boats behind us. We thought wow, we're a real airline now. The whole office came out and rolled up their sleeves to help wash and wax the first DC-3, even the cipher girls from accounting got in on the act. But the DC-3s don't remember very well—they just made us like every other airline. The Strannies made us unique. Of all the planes we flew, they had the most personality by far.

We have all read about oldtime sailors and their many beliefs and superstitions, and how almost human personalities were associated with the ships they sailed in. Some ships were "lucky" ships and came through the worst storms and the worst battles unscathed. Other ships were not so lucky and were forever becoming involved in one or another type of marine disaster, not necessarily related to the competence of the crews involved. There were some ships cursed with such bad reputations that owners had difficulty in getting crews to man them.

This sort of thing is not normally associated with aircraft, probably because of the very nature of their operation, and the frequency with which the flight crews are changed. A modern aircraft flies almost continuously, but changes crews every few hours. It may fly halfway around the world and back, and take aboard six or eight different crews in the process. Things just don't have a chance to get that personal any more.

But in the early days of QCA it was different. Any given aircraft would be doing well to average four to five flying hours a day. The rest of the time it would be in the shop being serviced, so it was not unusual for an aircraft to have only one pilot for long periods of time. Pilots would come to have their favourite aircraft, to which they preferred to be assigned. This probably was particularly the case with the Stranraers, since they flew only in daylight, and their entire flight crew of captain, first officer and flight engineer would stay with the machine for weeks at a time. So individual Stranraers soon took on personalities that stayed with them. Most were good — some not so good. CF-BYJ was a case in point.

I seem to recall that CF-BYJ was the third of all the Stranraers we bought. We originally had two — CF-BYI and CF-BYL — which we were using on the run to the Queen Charlotte Islands and Prince Rupert.

When BYL was lost on the mercy flight from Stewart it left us short, so we got BYJ and immediately flew her back to Montreal for conversion to a passenger plane. It was on this, her first flight, that she got into her first real trouble. On arriving in Montreal and attempting to land in the St. Lawrence River at the Vickers Aircraft plant, just above the Lachine Rapids, the river was very low with broken ice packs and the pilot, in trying to avoid the ice, taxied her right onto a submerged rocky reef, taking part of the bottom of the hull out and sinking her in shallow water. What was to be a fairly simple conversion now developed into a two-month repair job. To start with, getting her through the ice pack and up on the shore was no simple task. A channel had to be cut through foot-thick ice with handsaws and, after getting her out, a large canvas and corrugated iron shelter had to be built over her so she could be thawed out and the hull rebuilt. Fortunately we were able to get our old friend Albert Racicot to do the job, and he had access to materials and spare parts. We also sent one of our crew-chiefs, Curly Nairn, and another mechanic from Vancouver to oversee and assist.

Finally she was ready to go, and just like new. The ice had almost gone from the river and one of the local types undertook to pilot her out into the main channel. "Carunch!" and she had a new hole in her bottom. The French pilot waved his arms around and claimed that the river bottom was all changed since the thaw! Who knows, maybe he was right, but it took another three weeks to get her ready the second time, and she finally arrived in Vancouver after taking the deep south route via Mississippi, Texas and California. Once out on the coast she was put right to work, and we bought two more Stranraers, BYM and BXO, just to make sure.

Incidentally BXO was the last of the Strannies left whole out of the lot. She was the only Strannie left whole in the whole world apparently, because in the 1970s when the Royal Air Force was looking for an intact copy to complete their collection of historical military aircraft at the Hendon Air Museum outside London, they came to BC and glommed BXO from a group of antique plane buffs

The QCA Anson fleet outside converted CPAL overhaul hangar.

Anson interior. Bud Lando with moustache, Johnny Hatch at controls.
Over: Strannies BYI and BXO docked at Sullivan Bay.

BYL floating with the aid of two logs at Ceepeecee.

Former QCA ship CF-BXO on display at Hendon Aircraft Museum, London, England.

and took her back to the old country in pieces. I visited her there in 1984, all painted in RAF colours as if she had never been built for the RCAF in Montreal or had a glorious career flying for QCA at all. I tried to bring this to the attention of the museum officials but they obviously didn't believe a word I said.

I felt badly about this, but I didn't say anything because who would understand? Here in the west we have a very tenuous sense of history, and even when we do end up staring it in the face we think it's just our own personal history, of no significance to the world. But it was surprising how many people came up to me and expressed indignation that the Brits had scooped the last living Stranraer on us and stuck it in their museum when it should have been in ours. Still, how could we really complain? A few of these flying history buffs, including the filmmaker and sometime TV weatherman Bob Fortune, had tried to raise enough money to keep BXO here and hadn't been able to. As recently as the 1970s the awareness that this old plane might represent something worth saving still hadn't really developed in BC. For a number of years it just seemed we had been too slow off the mark and lost our chance to preserve what the Strannie represented, until BYJ changed that. But I'm getting ahead of my story.

After her second sinking in Montreal, BYJ got back home and flew the line relatively routinely for a few months. I say relatively because this plane couldn't seem to go two weeks together without throwing something at us. One hot summer day in 1947 the boys were welding up a crack in her exhaust ring just outside our rented TCA hangar on Sea Island when a spark went the wrong way and something started burning. In the time it took them to grab the nearest fire extinguisher and hose down the flames, three wing panels had been destroyed, at a replacement cost of twelve thousand five hundred dollars. The incident also convinced TCA we weren't the type of neigbour they wished to have sharing their classy digs and they threw us out, leaving us without adequate covered work space for several months. It was perhaps around this time BYJ got her nickname. We had been in the habit of naming all our aircraft "Something-or-other Queen." The "Skeena Queen," the "Nimpkish Queen," the "Haida Queen" and so on. BYJ spent so much time in maintenance she was dubbed the "Hangar Queen."

Then she encountered her third sinking, and this was a dilly.

It started as a regular scheduled flight from Vancouver to Zeballos, via Gold River, Tahsis and CeePeeCee. Andy Anderson was the pilot with Bill Oliver co-pilot and "Boost" Coulombe flight engineer, and there was one passenger aboard when she came in for a landing at CeePeeCee on a grey February morning, with reduced visibility in sleet and rain. The approach was normal, but right after touchdown she struck a partly submerged hemlock deadhead, with the top barely breaking water, ripping a five-foot gash in the bottom of the hull. The flight engineer saw what had happened and yelled at Andy to pour on the coal and keep her steaming for the closest beach. Andy opened the taps and bumped her as far as she would go up on the clam beach in front of the cannery, where she settled immediately in about five feet of water.

Andy said that locals appeared from behind every stump. Just before going up on the mud, they had passed a small boat with two people in it, and had practically swamped it with the waves they made. It turned out that the boat contained our one outbound passenger, who had engaged an Indian dugout to take him out to the aircraft. When the plane roared right on by, the anxious passenger thought he was going to miss the flight and urged the oarsman to "Follow that plane." When the plane stopped, the passenger jumped right in through the open door with his suitcase and landed up to his waist in water inside the hull.

"It's okay, Mac," Andy said, "we're not going anywhere right now!"

Our inbound passenger didn't fare much better. He had been standing up in the aisle when she touched down, and when she hit the deadhead, a great stream of water and wood chips hit him square in the chest, filling his eyes and mouth and soaking him to the skin, but he took it all as part of his day's work. He was an official log scaler, employed by the BC Forest Service, on his way to the camp at Tahsis to scale some timber. According to the story our crew told, he just mopped his face off and spat out the water and wood chips, but he stopped to chew on a particularly large sliver of hemlock. Then he spat it out and shook his head, commenting to himself, "Just Number Two Common. Must be out of the cull boom."

The first word of the mishap reached us in Vancouver by radio, but the message was, of necessity, very meagre and completely lacking in detail. It just said BYJ suffered slight damage at CeePeeCee and suggested that our superintendent of maintenance, Charlie Banting, come over and take charge. We should have been better informed, but it was by now company policy to say as little as possible on the radio, since too many people were listening, and if

Carter Guest got hold of it there would be no end of complications. As it was, the mere suggestion of needing Charlie Banting confirmed our suspicion that all was not well.

It was quickly decided that John Hatch would fly over with a Norseman, taking Charlie Banting, another mechanic and lots of tools; Johnny thought I should go too—just in case there were any higher politics involved. We arrived the same afternoon. It was a dismal sight. The tide had come in to cover the entire hull and cockpit. Only the wings and tail structure were above water. There was still intermittent rain and snow.

Our first problem was to provide enough flotation to get the aircraft up on the beach as far as we could so we could work on the hull when the tide dropped. There was a lot of talk about air bladders, which we didn't have, and oil drums, which are hard to harness. My contribution at this point was to visit the local logging camp and arrange to borrow two large spruce logs, which we got a fisherman to tow into the bay for us. Then we borrowed a lot of heavy rope from the fishing company. When the tide went down we floated these logs under the lower wings, one on each side, and then passed the rope over the logs and under the hull many times to make a sort of rope cradle. When the tide came in she gently raised off the bottom and we were able to move her ashore so that on the next low tide she was high and dry on the beach. We of course released the rope slings and floated the logs out as the tide fell. Low tide was around midnight, and it gave us a total of about three hours to repair the hole in the hull. It was about five feet long and a foot wide.

At this critical moment in history we were joined by a very useful chap who was travelling for the Standard Oil Company, doing repair and maintenance work on their marine refuelling stations. He was a marine engineer by trade and had a wealth of experience in ship salvage work. He offered his help and we decided to go along with his suggestions. It was very simple: plug the hole with concrete. We didn't think it would set in time, but he had an answer for that. You mix caustic soda with Portland cement and use boiling water, and this would set up in half an hour, he said. The canners had both the caustic soda and the Portland cement, so we were in business. All we had to do was dig our way under the bottom of the hull and screw on a large sheet of galvanized iron to act as a retaining form while the concrete set. It all worked just like the man said, and the morning's high tide floated her off high and handsome. It never leaked a drop.

Telling about it now makes it sound so easy, but the undertaking had its moments. Bear in mind that the actual patching job was carried out at night with the aid of a borrowed Coleman gas lantern, and with a chilling mixture of rain and sleet coming down. Charlie Banting came from Manitoba and he didn't mind cold weather or snow, but he hated rain. He was still wearing his prairie-style overcoat. It was very thick and heavy, came down to his ankles, and the collar turned up around his ears. But it was not rainproof, and during that night it must have taken on about twenty pounds of water. It sagged right to the ground like a tent, and Charlie could hardly move in it. The hem bulged out with water. Charlie was standing holding the lantern while Boost was lying on his back under the hull in a trench we had dug in the clam bed, working with a hand drill, fastening this large piece of galvanized iron to the hull with PK screws.

The trench was only about a foot deep and had filled with water, and Boost hardly had room to turn under there. Then I noticed something peculiar. Every now and again the steady stream of water running into the trench would suddenly swell to a subdued gush, and I noticed bits of paper mixed in with the dirty water. I began tracing this stream uphill with my eye and a few feet up the beach I spotted the partly-covered outlet of the cannery building sewer. Every time someone flushed a toilet, poor Boost got a soaking. It was running in the neck of his Mackinaw shirt and out his pant legs.

I pointed this out to Charlie and commented that we couldn't have picked out a worse spot on the whole beach. He looked things over slowly, with the rain running off the end of his nose, and then looked at me over the top of his glasses.

"Wahl," he drawled, "you cain't have everything *absolutely* perfect."

In the morning she was floating like a cork — no water coming in. All seats, floor boards, and lining had been stripped out of her. All the wiring in the hull and all the instruments in the cockpit had been under water and ruined, but fortunately the engines in a Stranraer are mounted in the upper wing and this had been ten feet above water, so they were undamaged. The only question was, how to start them? The wiring that ran the electric starters was out of commission and there was no way to crank the engines. This was no problem for our Charlie. He moored the Strannie alongside a log float. Then he got a fisherman's gumboot and made a fifty-foot length of rope fast to it. He had someone climb up and slip the boot over the upper propeller blade of the far side engine. Then he commandeered four enthusiastic locals and got them on the other end of the rope. He signalled to Andy for prime, and "Switch-On," yelled "PULL YOU BASTARDS!" and they did. The old engine fired first pull, the process was repeated with the near side engine and Andy was on his way back to the hangar and a third new bottom for BYJ.

Up to this point no one had been hurt in any of BYJ's tantrums, but a very definite pattern was emerging. In all three incidents sinking was involved, and in all cases snowy, icy weather was a factor. The fourth and final episode was true to form as far as sinking and snowy weather was concerned, but this time there was no happy ending. In retrospect, it was just as though old BYJ was trying to tell us something and we wouldn't listen, so finally she threw the book at us.

It was Christmas Eve, 1949; the weather — reduced visibility in wet snow. Pilot Bill Peters was heading north from Sullivan Bay with four passengers. While landing in the little bay in Belize Inlet–Allison Sound where Oscar Johnson's floating logging camp was then located, something went wrong: just after touchdown as Bill started his "round out" at about seventy-five miles per hour, the nose dug in, sheered sharply to the right and the big machine executed a complete cartwheel. The whole nose section of the hull broke off. Bill got thrown through a hatch but was fouled in the wreckage and dragged part way to the bottom before struggling free. Co-pilot Jack Steele and flight engineer Sig Hubenig were momentarily stunned but managed to recover in time to swim out through the break in the hull. Two passengers, Gordon Campbell and Gordon Squarebriggs, were able to save themselves but the other two, Ralph McBride and John Buckley, were pulled down and drowned. The ship sank in three minutes with only the very tip of the upper wing showing. Oscar and his crew were able to get a line around it, and dragged it to the surface with a boom-winch to remove the bodies of Buckley and McBride, who were still strapped into their seats. We theorized that they might have got out if their safety belts had not been fastened, and for this reason we generally advised seaplane and flying-boat passengers to leave their belts unfastened during take-off and landing, but to secure them in flight in case of turbulent air. After this, Oscar lifted the wreck out with a logging donkey and A-frame, and piled it on the beach.

I am not a superstitious person, but by the time I was finished with

BYJ I had almost become one. Her ghost lay quiet for thirty-five years.

In September, 1981 I was taking an extended cruise up the BC coast with three friends, all of whom were already previously well acquainted with the coast but, like me, always willing to see more, so this was one of our "Voyages of Adventure and Discovery" in the good ship *Blithe Spirit*, which I had purchased after I left aviation. None of the other three had ever heard of BYJ, and I had no idea that there was anything left to see or on what part of the sixteen hundred miles of shoreline in the Seymour Inlet complex it might be.

We had been looking for a suitable spot to set our prawn traps but after about twenty miles it was still too deep, so we decided to find a spot to anchor for the night. The Seymour Inlet group had not yet been charted, so we were quite on our own. It was about 10 P.M., and after crawling along with both echo-sounders running we found our way into a narrow entrance. We felt our way through several right-angle turns, located a secluded bit of channel in which we were able to get our anchor down in a convenient twelve fathoms of water, and secured for the night. In the process of positioning the ship I had been sweeping the shoreline with the searchlight, and in doing so had picked up an unidentified object on the beach that might well have been the wreckage of an old building or something. In the morning I put the binoculars on this object and my spine just turned to ice. I launched the dinghy and rowed ashore before the others were up, and there it was. The remains of CF-BYJ, covered with moss and seaweed, but with part of the QCA speed-line and name still readable in black and yellow on the side of the hull. The spectre of Christmas Eve 1949 back to haunt me! Scavengers had been at it but, Belize Inlet being uncharted, it is not frequented by either pleasure boats or commercial fishermen, and the wreckage was still recognizable as a Stranraer. Some of the parts were remarkably well preserved.

In 1983 a tug, scow and crane picked up all that was left and brought it down to the Canadian Museum of Flight in Surrey, with the intention of rebuilding her to once again look like CF-BYJ. I approve entirely, as long as no one tries to fly her when it's snowing.

Into Wider Skies

JOHNNY HATCH REMINDS ME that the first time we needed to have someone fly to Ottawa, I forget what this was over, to pay court to some mandarin or other, he said to me, "Who's going to do this?"

"I don't know. Who do you think?"

"I think you should go," he said. "You're the president."

"John," I am supposed to have replied, "I'll never cross the Rocky Mountains. Never."

That might even be true. It's hard to realize it now, but up to the time I was about forty I'd never left the coast and I never intended to. Boy, did that ever change in a hurry. You don't do anything in this country, least of all run a highly regulated business like an airline, without you end up spending a lot of your time on bended knee in that godforsaken outpost called Ottawa.

Trans Canada Airlines was offering the only service across the Rockies in those days, flying fourteen-passenger Lockheed Lodestars to Calgary and Lethbridge, then DC-3s the rest of the way—Regina, Winnipeg, North Bay, Ottawa. The first time I went, in 1947, it took two days via Kapuskasing, but a more normal time would be about eleven hours. Later on the cross-country run they got something they called the "North Star," which was essentially a DC-4 with Rolls Royce Merlin engines in it, a miserable mixup of an aeroplane. Terribly noisy and very unreliable. In fact our people wouldn't fly them. We used to go to the States and fly United into Toronto just to be in a safe aeroplane.

The thing that made me almost a dual citizen of Ottawa was the postal subsidy. Soon after we got flying regularly we began to be prevailed upon to carry letters and packages up and down the coast for people. Unofficially, of course. We had our own mailbags the pilot would take along with him and we even had our own little postmark, "Courtesy of QCA." We were soon delivering almost as much mail as the Union Steamships, but they were getting $375,000 a year for their trouble and we were only getting smiles.

We enjoyed being nice guys, but since we were chronically short of liquid capital, it wasn't long before someone got the bright idea we ought to apply to Ottawa and get us our own $375,000. All the other airlines must be getting piles of money for mail, we thought. Traditionally in the US and Canada the government had subsidized the airline industry by providing a very large part of scheduled carriers' revenue in the form of mail pay, and this was still going on, we thought. We didn't really know. We didn't know very much about anything, except how to get over the Alberni hump in all sorts of weather and how to keep a Dragon Rapide running a regular schedule ten years after it had personally decided to retire. To gain any of this broader knowledge, it always turned out you had to make a trip to Ottawa.

The first three or four times I went to Ottawa in search of QCA's share of post office gold I couldn't even find the right guy to ask. This led us to make another breakthrough onto the world stage, which was to join up with AITA. AITA stood for Air Industries and Transport Association, and everybody who was anything in Canadian aviation came to their general meetings, held once a year, usually in and around Ottawa. Later the Air Industries, which included equipment manufacturers and dealers, went off on their own, leaving the airline operators alone under the name Air Transport Association of Canada (ATAC), but at this stage we were coming together as one big group. The first time I went I took Mike de Blicquy with me. He was our assistant operations manager at the time, a remarkable man, an oldtime flyer of Belgian extraction and a very knowledgeable guy about flying. He was also a Belgian count, which he didn't like anyone to know, but it gave him a well-bred air — at least I imagined it did. Venturing out from under my west coast rock to face the national scene for the first time, I felt I needed somebody like Mike to hide behind.

The 1949 meeting was being held at the Gray Rocks Inn at St. Jovite near Ottawa. It took three days. A man named Cotterell from TCA was president, but the organization was run by its executive director, Bob Redmayne, and his assistant, Hal Suddes. I met a lot of people there and made quite a nuisance of myself picking people's brains. It was a feast of inside information the like of which I'd never come across. We did a kind of survey to find out what other companies were paying their pilots and maintenance men to see how we stacked up.

Milt Ashton, of Central Northern Airways in Winnipeg, paid his pilots $185 a month plus three cents a mile. The pilots got a ten-dollar raise for each year of seniority and topped out at $280 plus the three cents. It worked out to about $400 a month take home pay. This was in line with our 1949 rates. Engineers got $200 a month. (We were paying our Superintendent of Maintenance at this time $250; ordinary engineers $150–$185.) Ashton had some sort of bonus plan that paid out of his operating profits at the end of the year. He didn't have any maximum working hours. They worked around the clock until the work was done. No overtime. Copp, of Maritime Central Airways (MCA), paid his pilots a flat salary varying from $350 to $500. They worked a forty-eight-hour week. And yes, they both got mail pay. We couldn't pry out of anyone precisely what their mail pay was because this was everyone's private little secret, but there was no doubt it formed a substantial part of all their incomes. Mike de Blicquy picked up a hot tip that TCA wanted to dump their Vancouver–Victoria run. They had to station aircraft out on the coast to service it and it lost money, so he was told. We sat up all one night in a state of high excitement scheming about how to snatch Victoria away for QCA before it fell to CPAL, but of course Air Canada was still quite happily flying the run in 1988.

I was too green to tell gossip from fact.

It was quite a thrill to be among my own kind, mixing with people who did nothing but live and breathe the kind of problems we had been toughing out all on our own all these years back at the coast. It was quite clear that the AITA was becoming a very powerful influence in Canadian aviation and it would be very much in our interests to stay close to them and use them as much as possible in connection with the post office, the DoT, the ATB and the whole Ottawa pack. In the hallways and hotel rooms between sessions a kind of Turkish market went on as all the gyppo operators tried to flog their used hardware at each other. Racicot was there with a Norseman on wheels, skis and floats for twenty thousand (too much). The Babb Company (Canada) Limited, represented by Ches Newhall, had an Anson fully equipped for IFR night flying for five thousand. I gave them our old original Waco, CF-AWK, for a down payment on it. Punch Dickins was around getting quite desperate to have us all stock up on his new Beavers, but none of us had the money. He started offering very extended terms, but we still couldn't see all that loot for a plane they were calling the "Baby Norseman." By our reckoning, the thirty-one thousand Dickins was asking ought to get at least twenty paying customers into the air, and the Beaver carried only six. Instead we arranged a deal whereby the Aero Club would buy one of his new Chipmunks and we would take the club's old Stinson for three thousand dollars. Now this was talking business!

Eventually the Beaver became available as ex-military craft from the US Air Force and that brought the price down a bit. They took over completely because they were a much more efficient aeroplane than the Norseman, and even though they were a lot smaller they'd pack just as much freight. They were far cheaper to operate. For power they had Pratt and Whitney Wasp Juniors, which was a more modern and less costly engine than the Series H Wasp in the Norseman. You could go fourteen hundred hours between overhauls, as opposed to five hundred on the Norseman. They were ideal for the small operator. It was easier to fly them at a profit.

After the AITA meeting ended, de Blicquy and I went down to Ottawa and visited the Air Transport Board—Alan Ferrier, John Baldwin, Romeo Vachon. They were very interested in our progress and had much advice to offer. Romeo Vachon said we should get rid of the Stranraers and replace them with large flying boats we could build ourselves. Build our own wings and fuselages and put in Pratt and Whitney engines and fly to Alaska and down into Washington and Oregon. He'd seen our shops and thought we could do it with no trouble. Vachon was an old experimenter from the early days of flight and he couldn't stop trying to dream up better flying machines. Alan Ferrier took a different tack. He said the secret was airborne radar, and especially colour television. Just how he planned to have us use colour television in the air escapes me now, but he liked it. And he said get out of fixed-wing aircraft as soon as you can. The future was going to be all helicopters. The best thing about them, the government didn't have to build airports. That was the advantage of flying boats, too. The government was all for flying boats. Then came Baldwin. He was altogether different. He wasn't an old flyer. He was a Rhodes scholar, quite young, and considerably more down to earth. But the other two shook us up.

They seemed a bit screwball. One thing they all agreed on, though, was that we should be getting substantial airmail pay on all our routes. They couldn't believe we were getting nothing for it up to then and implied it was rather silly of us—they made me promise to go right over to the office of the Postmaster General, George Herring, so this amusing anomaly could be fixed up before I left town.

Following the ATB I went over to the Department of Transport. There was always a bit of an unclear division between the authority of the ATB and the DoT's Civil Aviation Division, with the ATB regulating routes and the DoT setting safety standards, monitoring operating procedures, granting operating licences, inspecting aircraft, operating airports, operating navigational aids, doing accident investigations, etcetera. The DoT was the larger department and the senior one, but the ATB had in many ways the most critical function in terms of directing the shape of the industry. In general we got along better with the ATB than we did with the DoT. I now went in to see the DoT controller of air services, Air Vice-Marshall Tom Cowley.

Cowley had a reputation as being a pretty crusty old devil but on this occasion at least I found him fairly approachable and we had quite a long talk. The big thing on his mind was Carter Guest. Carter Guest was still in charge of the DoT office in Vancouver and as usual we were getting very poor cooperation from him. He'd gone on giving us more and more trouble as we grew, and all the other operators found him just as unreasonable as we did and assailed Cowley with a steady stream of letters complaining about this backward, narrow-minded old autocrat who ruled flying on the coast like a private fiefdom. Cowley asked me point blank if he should fire Guest. I said I didn't know if he should be fired, but it would be a big help if he changed his attitude. He was causing us a lot of hangups and costing us a lot of money.

While I was in the DoT office I got talking to someone in the radio division named Des Murphy. I was surprised to learn this man had once worked as a logging donkey engineer, as I had in my early twenties, and one of the places he had worked was Theodosia Arm, which was the actual place I first broke in, although he had been at the big Merrill, Ring and Moore camp while I worked across the inlet for Palmer Brothers. How he got from there into radio and in Ottawa, I don't know. Well, how did I?

On my way home I made a detour through New York to check out a Norseman I heard was for sale in Secaucus, New Jersey. I got to

Manhattan and booked in at the Biltmore Hotel. The entry in my trip journal consists of one word: WOW! I looked up a secretary who was on holiday down there and together we wandered around and saw the lights of Broadway. There was one billboard in particular I couldn't get over, for Lucky Strike cigarettes. It had a man's face about a hundred feet high and he would would blow a big string of smoke rings out over the street. It made an awful roaring noise, and at night they had it lit up so you could see this smoke magically appear. I was very taken with it and stood around with all the other yokels, gazing up while the townies pushed past muttering insults.

The next day I tried to hire a cab to take me to Secaucus. As far as I could see, it wasn't really very far, just across on the other side of the river, about like catching a cab from downtown Vancouver to the airport. But not one of these guys would touch it. "See what? See-kaw-kiss? Nevah hoid of it Mac. You sure?" I pointed across the Hudson, which is really quite a small river compared to the Fraser. "Oh, Noo Joisy! Noo Joisy, you shoulda said that." Well, Noo Joisy was out of the question. They were Noo Yawk cabbies and if I wanted to go someplace in Noo Yawk, they'd take me. If I was going to be such a moron as to venture into that wasteland beyond the River, well, that was my problem. They didn't want to talk to me. They rolled up their windows. I talked to various people and they just shook their heads. "New Jersey? You want to go to New Jersey? What for?" You'd think I was proposing an overland safari to Tierra del Fuego. In any event the only way there was to go down to the Port Authority and take a bus—but I would be on my own. They wouldn't be held responsible.

Well. I'd never seen such a setup. This Port Authority bus terminal had floor upon floor upon floor of different levels of buses, all taking off for different parts of the country. Very confusing. Nobody would give me a straight answer about anything, but I eventually landed up on a bus that claimed to be going through Secaucus. We whipped out of the terminal and through a tunnel under the river, and soon after emerging on the other side found ourselves rolling through pleasant farmland. It was such a relief from the tense atmosphere and choking fumes of Manhattan you had to wonder what kept people over there. I sat directly behind the driver and I said to him, "I'm a stranger here and I want to get off at. Secaucus." I thought I heard him grunt. It was some sort of noise, and not a friendly one, so I didn't feel I should risk bringing the matter up again. Better to sit and fret. He did announce the stops, but in such a surly, grunting fashion I could only guess what he was saying. We raced along, clickety-clickety-clickety-click. We passed one place after another until finally I recognized the name of a small town *away* down state. I tapped the driver on the shoulder and said, "When do we get to Secaucus?"

"Oh, you mean *Sekoikus*," he said. "Hell, mister, we passed Sekoikus half an hour back."

"I didn't hear you announce Secaucus," I said.

He fixed me with a withering stare.

"There ain't nothin to announce about Sekoikus," he said.

He then wheeled over to the side of the road, slammed on the brakes and said, "You can git orf here." I didn't have the proper ticket to be on his bus any farther.

"How will I get to Secaucus?" I asked. Quite impatiently he explained that if I stood on the roadside and flagged down the green bus coming the other way, I'd get to Sekoikus, which he could now see was where I belonged anyway. He didn't say that in so many words, but he managed to make it very plain to everyone else on the bus. Whom I could tell all agreed with him.

A green bus did eventually come, and I did eventually get to my destination, which was the office of Mr. Leo Van de Wal at the Dawn Patrol Airport. Dawn Patrol Airport was nothing more than a farmer's field with a few old wrecks of aircraft sitting around, including a Mark VI Norseman, Serial Number 427. It was painted black except under the wings and fuselage, which were painted an ivory colour. It was in horrible shape. I later learned that the reason for the fancy paintjob was that this plane had been in the business of flying wetbacks over the Mexican border during harvest time. From below it blended in with the clouds and from above it blended in with the pigfarms. However, this haywire-looking character who owned it had a set of floats for it and claimed he could get an export certificate. The price on floats was seventy-five hundred dollars, about half the going rate. That made it suddenly look a lot better. I gave him a cheque for five hundred and phoned Vancouver. It was arranged that Johnny Hatch would come out and get it, but from here on it becomes Johnny's story.

> Spils phoned and told me he'd located this aeroplane and made a deal on it.
> "I want you to come and get it," he said. I was in Vancouver and he was in New York.

"Okay, when I get it what do I do with it?"

He said, "You take it to Vancouver."

Of course.

You ninny.

"You realize," I said, "that every lake and landing area for a seaplane between where I am and where you are is frozen solid?"

This was February. There was a long silence at the other end of the phone. Finally Spils said, "You're the operations manager. You figure it out."

"Well," I said, "I'll come and get it and I'll take it home, but it'll be the long way around. I'll have to take it south across the Gulf of Mexico and up the Rio Grande." I could have asked him for the money to go buy or rent a set of landing gear so it could be flown back direct, but going this way seemed less bother.

"Just don't be too long," he said. "We need you back there." This was the kind thing that gave all of us in ops a special place in our hearts for Spils. Give you an impossible mission and lousy equipment to do it with, then tell you to hurry it up.

As it was, I assured him it wouldn't take much over a week's time. The plane had a range of two thousand miles, and flying eight hours would give me an easy thousand miles a day. The continent is only four thousand miles across, and even doubling that for detours, I couldn't see it going over eight, nine days.

Part of the deal was this aeroplane had to be flying and certified airworthy. When I got there, it wasn't. The wings were off it. I called Spils and told him there was a problem. "Oh," he said, "There's a bunch of stuff for sale down in Florida. Go down there and see if you can find us some R13-40 Wasps and some Edo floats."

I went back up to Secaucus after about ten days and the wings were on but the aircraft still wasn't certified. When we finally got ready to fly it was well into week three of the trip I had budgeted seven to ten days away from the office for. I'd never flown in

that part of the world, so I went to the Civil Aviation
Authority and got the seaplane register for the
United States. It listed an impressive network of
seaplane bases all along my route. I started out
aiming from one seaplane base to the next. The first
one was closed for the season. So was the next. So
was the next. In the wintertime they just didn't do
seaplane flying. This meant any time I stopped it was
an automatic overnight, which stretched the trip out
considerably.

The first night I spent in Florida, the second night
just outside New Orleans, working my way down the
coast following a series of sloughs and canals used
by the sports boating people. You want to stay pretty
close to water in one of those things, particularly
when you don't know the aeroplane very well and
you're a little suspicious of its previous owner. By
the time I left New Orleans I realized I was going to
have to do without regular seaplane base facilities — I
couldn't just land, taxi up to a dock and expect to be
refuelled. So I had to figure out where I'd be the next
day and wire ahead to arrange for somebody to meet
me with a tank truck. That way I got to Fort Worth,
Texas and I was going to land next near El Paso at a
lake that was four miles long — I had given up the
idea of flying along the Rio Grande because the
United States had gone through a seven-year
drought at this point and there was no water in the
Rio Grande. On my map it showed up as a bold blue
stripe, but from the air it was just a string of
sandbars. This lake which was shown on the map as
four miles long had just dried up to nothing. I flew
over it and there was just nowhere you could put a
floatplane down. So I got out my map and found the
nearest really big piece of water. There was
something in New Mexico called the Elephant Butte
Reservoir that looked like it was sixty miles long, so
I flew back there. If that's dried up, we're really in
trouble, I thought. I had just enough fuel to make it
one way.

I found it and flew over it. The water level was

something like four hundred feet below normal. It was just chopped up into a whole bunch of little lakes in the bottom of this miniature Grand Canyon. The longest stretch of open water was two-and-a-half miles.

The place I was tied up was a sports fisherman's dock. After taking on fuel the next morning I got everything fired up, cast off, turned away from the dock, put on a little power and the aeroplane just went "CRRRUNCH!" Stopped dead. I got out on the floats, lifted the inspection covers on the float that was aground, and two compartments out of seven were leaking. I hollered for a boat and these guys came out and said, "We're sorry, we should have told you about those rocks!" This water was so muddy if you stuck your finger in you couldn't see the end of it.

I decided to go back to Fort Worth. When I refuelled there on the way out I noticed an abandoned naval air station with a huge wooden ramp right down into the water. I thought, I'll go back and I'll run it up on that ramp, jack the plane up and get somebody out to repair the damage.

We got some power boats around and got the plane off. It went down heavy on one side. I went out and struggled this thing into the air with about half a ton of water in the floats. They said it left a cloud of water behind for about a mile after I took off, like a water-bomber. I got to Fort Worth, found the ramp, slithered up out of the water and shut 'er down. Then I started phoning people. I was told there was a good outfit that would come out and do a repair, so I phoned them and this big drawling Texan comes out, takes a look and says, "We can do 'er!" He had a war surplus crane he'd bring out, raise the plane up on blocks, and replace the float panels right where she sat.

I went into the hotel and parked in front of the phone. I picked up the receiver but couldn't think of what to tell Spils, so I just dictated a wire, "Send

$1000." I checked in, went to my room and waited for the inevitable phonecall. It came right on schedule. "What's the story, John?" Spils asked. I told him the long, sad story. I said I should be on my way in a day or two. He agreed to part with the thousand.

The next morning bright and early the big Texan brings out his war surplus machine. It's got a big, welded derrick on it and he swings the derrick out over the plane, puts cables around, lifts the plane up, and the arm on the derrick just goes "CHOINGG!" — breaks and falls right into the aeroplane. The whole side of the fuselage is caved in. One of the longerons, the main structural members of the aeroplane, is knocked galley west.

I was speechless. I was walking around, looking up at the sky, looking down at the ground, thinking, "Why me, God? Why me?"

This Texan in his slow manner comes up to me, claps a hand on my shoulder and says, "You know, you look like you got all the troubles in the world on your back. And I can see you have got a fair slug of 'em, but I tell you what I'm goina do for you. I'm goina fix those floats. And I'm goina fix that aeroplane. And I'm goina get it certified for you. And I feel so damn badly about it, it isn't goina cost you a dime!"

And he went ahead and did it. He brought out welders, he peeled back the fabric and he packed wet asbestos fibre around the frame so he could cut and weld the steel members without setting everything on fire, and inside of forty-eight hours he had me back in service.

I set out west across New Mexico, figuring to make the coast in two days.

In Canada there were certain prohibited areas, like the old BC Penitentiary. On the map it said "Prohibited Area" and you stayed above five thousand feet when you were flying over it. I was steaming across New Mexico after I got back

together and I see a place ahead that's marked
"Prohibited Area" so I go up to nine or ten thousand
feet and cut right across it.

All of sudden I've got an aircraft out the port side
and I've got an aircraft out the starboard side and
they're both jets and they're staggering along trying
to stay close to me—they've got the gear down,
they've got the flaps down, I'm doing my usual 90
mph and that's practically their stall speed. I can see
the pilots making frantic signals, "Down, down,
down."

If they wanted me to land they were out of luck
because as far as you could see in any direction there
was nothing but real sandy desert. I get out the map
again and the name of this place is White Sands,
New Mexico. That meant nothing to me at the time
and it wasn't marked as anything special, but of
course, as we know now, it was where all the
top-secret atomic testing was going on all through
the forties and fifties. It was not only prohibited to
small planes, it was off limits to everything in the
sky, and it was damned dangerous to get anywhere
within a hundred miles of. That tattered old
Norseman must have had sirens going off in
underground bunkers all over the free world.

Eventually they got tired of trying to hold their
high-speed interceptors in the air near me—they
would have picked out the Canadian registration—
CF-GRU—which I had fortunately had painted on
the side before I left Secaucus—and they let me
stumble harmlessly along on my way.

I overnighted on Lake Havasu that night and in
Frisco the next. Coos Bay, Oregon was the first, last
and only place on the whole trip where a facility that
was shown on the map as a seaplane base actually
turned out to be an open, functioning base where
you could get gas for the asking and file your flight
plan without having to explain what one was.

When I finally got back to the Vancouver Airport
and reported to the office Spils said, "Hmph! If I'd
know it was going to be that much trouble I'd have
shipped you a set of wheels."

Worst of all, everybody who glanced over this plane to which I'd just given five weeks of my life was quite outspoken in saying we all would have been further ahead if I'd left it in Secaucus. Harry Lewis, our aeronautical engineer, confirmed it. He ordered the aeroplane into the hangar and stripped it right down to the bones. They rebuilt it from the ground up. By the time maintenance was finished we were probably over the twelve thousand dollars or so it would have cost us to get a decent Norseman Mark IV just down the block.

I began our long campaign for mail pay when I was passing through Ottawa for my first AITA meeting in 1950. I went to the parliament buildings and met Jim Sinclair, the Minister of Fisheries from BC, and Jack Gibson, the MP for Alberni riding. Both of them wanted to see us get an airmail contract, both of them wanted to see airmail service on the coast expanded, both of them wanted to see us get properly paid for it. They were very aware that the old Union Steamship service which had been the coast's main link with Vancouver since 1900 was during this period in a state of rapid shrinkage, and it made sense to them that as the USS mail service was cut back, their mail contracts should be transferred to QCA with no decrease in pay. When I went in to see them the next day they had been to see Mayhew, the minister, and found him enthusiastic. They were all three going to see George Herring, the postmaster general, on QCA's behalf the next week. When I got back to Vancouver I told the boys our postal money was on the way.

But time passed and nothing happened. I was puzzled. I called up Sinclair again. He was still working on it. He reminded me nothing happened quickly in Ottawa. Eventually we got a letter from the post office notifying us we would be granted a contract for twice-weekly mail delivery on our West Coast route only, calling at Tahsis, Ceepeecee, Zeballos and Chamiss Bay. The pay would be thirty-five dollars per week. We were furious. We considered refusing it, but the directors decided inasmuch as we were presently carrying the mail as a courtesy no harm could be done by accepting. By the time the next year's AITA meeting came around we were still in substantially the same place we'd been the year before.

I had sat around respectfully listening during my first AITA meeting, but the next year Johnny Hatch and I put our heads together and decided to bring forward a proposal to straighten the government out on this mail pay thing. We said enough of this, let's demand the government pay us all for the mail we deliver on every route, and they should pay us at the same rates we charge our normal customers for hauling air freight. For us, this would mean a many-fold increase over what we were presently getting. I was quite forceful. We were met with stares of blank disbelief. Then, one after another, the other operators from across the country stood up and denounced our plan. One would cite one cockeyed reason, the next another. What nobody said but we finally came to realize was that they were all being paid far *in excess* of standard air freight rates already. They didn't want the post office to get the idea they would ever settle for mere freight rates. When it came to mail pay they really didn't want to rock the boat. As for us, we had their sympathy and they all privately felt we should go have a talk to somebody in government.

I should mention that on most of these trips to Eastern Canada a great deal of my time was taken up interviewing prospective customers for Spilsbury and Hepburn radio equipment — government departments, other aircraft operators, etcetera. Besides rustling up new customers, there was already quite a lot of our equipment out in the field and a lot of followup work to be done on that. It frequently took half of my time on a given trip and was quite successful. By the time I was done we were selling radio equipment all across Canada.

After the meeting was over Johnny left and I went in to meet the postmaster general, George Herring. I went through the procedure of asking for more appropriate mail pay rates and to carry the mail more frequently over the routes that we were serving daily. We were still only hauling mail twice a week on the West Coast and people wanted it every day. Herring wasn't so keen on adding to the frequency, but he thought maybe the time had come to add more delivery points. He suggested calls of perhaps once a week at fifteen dollars per landing and takeoff. But nothing for actual transport. As these were mostly what we called "flag stops" — places we only called on if we had word there were passengers waiting — Herring's plan would leave us in the position of having to fly out of our way and make a stop even if we didn't have a passenger, for fifteen dollars — which wouldn't begin to cover our cost. In general Herring agreed we were getting peanuts for our mail service, but said he couldn't do anything about that as postmaster general. He urged me to keep hammering away at the MPs, saying only if they raised enough stink about it would it eventually get addressed.

I took this news directly to Skeena MP George Applewhaite. If he wanted airmail service to Ocean Falls and Alice Arm and the Queen Charlotte Islands, and Sinclair wanted more service down the south coast, along with Jack Gibson on the west coast of Vancouver Island, all these people would have to get their shoulder to the wheel. I would pass along abstracts of our various submissions and we would get the local Boards of Trade busy in all these places. Before long we had a real program going, soliciting help from all the communities involved, hoping to impress the post office, who said they'd love to do it but couldn't without public support. But still nothing happened in Ottawa.

After the 1950 AITA meeting ended I was joined by Hugh Mann, who was one of our senior pilots. The idea was that we would set out together for Europe to look for aircraft there. Hugh wanted to accompany me at his own expense, just for a holiday, which was very convenient. We flew first to Gander, where we inspected four Norsemen owned by Maritime Central Airways. They were in very poor shape, and when we looked at MCA's hangar we saw why. Their service was very limited. This surprised me. MCA owner Carl Burke was quite a veteran in the business and had a lot of respect in AITA circles. He never seemed to have any of the struggles with the DOC and the ATB that we did, and he got the highest ratio of mail pay to passenger revenue of any airline in the country. The sign over his hangar listed his regular stops as Charlotteown PEI, Summerside PEI, Moncton NB, Saint John NB, Fredericton NB, New Glasgow NS, Halifax NS, Sydney NS, Magdalen Islands PQ and the French Islands of Saint Pierre and Miquelon. It made an impressive route map to my eye; I could hardly believe he was behind QCA in annual revenue miles flown, but he was. We picked out the best-looking Norseman of the bunch, CF-GRQ, and Johnny came back out to ferry it home. Later we bought more. I think we got all of them in the end.

The real reason we were going to Europe was to look at some Sandringham Flying Boats offered for sale by the Scandinavian airline, SAS. This was the first time I'd ventured back to Europe since stopping their briefly in 1905–06 for the purpose of being born, and everybody at the office was quite excited on my behalf. Lando especially was concerned that I not do anything that might blacken Canada's good reputation in the mother country. He gave me a long lecture on the various counts on which my primitive Savary Island ways might want correcting. The only particular I can now recall was that I was not under any circumstances to leave my shoes outside the door of my hotel room at night, notwithstanding my objection that this was standard practice at the Alcazar in Vancouver. According to Lando, there were so many ways a colonial could go wrong in England it seemed a virtual minefield of disgrace. He managed to get me quite worried about it and made both Hugh and me read a book he loaned us with all the proper methods of tipping and ordering etcetera. The book confirmed Lando on the matter of the shoes, saying the one and only proper thing to do was to summon the hotel valet to take them decently away to an odour-proof locker in the nether reaches of the building.

We hadn't made any arrangements to get across the Atlantic, but while we were in Gander a Pan American DC-6 freighter arrived empty, deadheading to London. We got talking to the pilots and they said, "Why don't you come along with us? Nobody else wants to ride with us. No charge." That looked like our kind of deal, so we jumped in.

We soon discovered why nobody else wanted to fly with them. This aeroplane had been chartered to bring a load of monkeys from Africa to the New York Zoo and the smell was just awful. We departed on the 8th of November, 1950 and arrived November 9 at Shannon, smelling just as bad as the plane. We refuelled and indulged our desperate thirst for a cup of fresh coffee, wishing immediately we hadn't. It was so terrible neither of us could finish it. Then we flew on to London and tried the coffee there. We couldn't finish that, either. Hugh booked us into the Normandy Hotel. It was a small hotel, very old-fashioned, but we were glad to get anywhere. We were dirty, we were tired, we were beat. We had a good night's sleep. We woke up in the morning and pulled up the blind and all we could see for miles were hundreds, thousands, millions, of chimney pots. We were just level with them. A forest of chimney pots. My first impression of London.

We went down, and even in a little hotel like this they had all this swag: the guy at the door had gold braid starting at his neck and going right down. Brass buttons, flaps and all the rest of it. Neither of us had travelled before and it was quite an eye opener all round.

We still smelled like two bull gorillas in rut and wanted to have a bath, but there was only one bathroom per floor. I guess there would be about four rooms per side, and there were four sides. In the very centre was the lift. The lift came up in an open cage, as I am told they still do in London. You could only get two people in the thing. To get to the bathroom from our room, you had to take quite a hike down the hall past the lift, but we hadn't brought any dressing gowns or anything. Lando had failed to prepare me for this. One way or another it ended up I was peeking out the door waiting until the can was empty so I could make a dash for it just in my shirt, bare-legged. Well, of course, just as I got halfway, the lift pulled up and inside was an old lady with a lapdog looking bug-eyed at me over her shoulder. I was about at the point of no return and decided to streak it anyway, got the bathroom door open, and fast slammed it shut. Inside was pitch black. Well, where do you find the switch? I feel all around, open the door a sliver to let some light in, and finally I find a whole set of switches. They were the push-button type. I pushed one. Nothing. I push the next one. Nothing again. I push another. Still no light. Finally I get looking around further afield and hidden on the wrong side of the door where you'd least expect is the light switch. So I got the lights on, locked the door, ran the bath, stripped my monkey-smelling underclothes onto the floor, where they lay radiating in a heap, and plunged in. About this time the waiter arrived to take my order. Then the valet. Then the maid. They all just walked in. They all had master keys. "Yes, sir? You rang, sir? Champagne, sir?" These were the buttons I'd pushed. They had their order pads at the ready, their clothes brushes out, razors, ready to go to work on me. I was mortified, but if there was anything that seemed the slightest bit out of the ordinary to them, they gave no sign. Actually, as the maid stepped over my monkey duds, which had originally exhibited a musty scent but had now advanced several octaves to a piercing aroma that worked on the brain like a high-pitched scream, I thought I detected a possible faint twitch of the nostrils. It may have worked to my advantage, because they were all very quick to exit once I blushingly explained the mistake.

The second morning we went down to breakfast, and at this time Britain was still very short of certain food items. We got menus from the desk and looked them over. They had porridge, but no bacon and eggs. I was alright because they had kippers, and I love kippers for breakfast. Hugh couldn't bear kippers; he couldn't bear even to think of someone eating them for breakfast. He wanted eggs. Well, the only eggs they had were plovers' eggs. They were tiny things like marbles and about a dollar each if you had the right family connections, and Hugh bullied his way into some of these. This was all arranged as we stood at the desk, then Hugh said alright, which way to the dining room? "Oh, the dining room is closed, sir." They sat us down in the lobby and pushed up a card table. We felt rather foolish, with people coming and going, but we put up with it. Hugh took out a paper, so I lit up my pipe. At this the fellow with the gold braid came running over and said, "Sir! Please extinguish your pipe! Pipes are *not* allowed in the lobby!" With that, something snapped in Hugh. He just stomped over to the desk and checked us out. Hugh does that sometimes. He just snaps. Oh, was he angry. And was I hungry!

We didn't know where we'd go after that, so we put ourselves at the mercy of a cab driver, and seeing that we were Canadians, he very sensibly took us to the Dorchester, which was the hotel Londoners more or less set aside for Canadians in those days. The first person we met when we got in the door was Bob Townley Jr., the nephew of one of the original members of the Savary Island Syndicate. We sat down for some breakfast and I discovered that he was also now in the flying business, and like us he was interested in seeing what de Havilland were doing while he was there. So we got down to their factory the next morning and had a look at this new de Havilland Dove we'd been hearing so much about. It was a pretty little two-engine passenger plane, but we agreed it was useless for BC coastal purposes.

What did catch our eye were two extraordinary-looking airliners sitting on the tarmac outside one of the large hangars. What was extraordinary about these machines, to us, was the fact they had no propellers. They were the world's first two jet airliners, the de Havilland Comet, copies number one and two. We rushed over and saw that one was warming up. Just then the well-known de Havilland test pilot, John Cunningham, walked up to us and said, "Would you like to come up for a flip?" So we clambered aboard Comet Number One and went up for about a two-hour spin over Europe. You couldn't see any of England. It was just covered with a blanket of smoke. The only sign of anything was here and there you'd see a trail of steam from a locomotive charging along, swelling over top of this solid layer of smog. We couldn't see anything else until we got up to forty-one thousand feet and began circling over France and the Mediterranean. It was the first time either of us had been up that high and we were very impressed. There were just crates to sit on, there was no proper finishing or lining, but the smoothness of the ride and the quietness was still very striking to anyone raised on propeller craft. Later, of course, the Comet Version One was found to have a structural flaw which caused it to blow apart at that kind of altitude, and after three crashed in rapid succession the rest were scrapped.

We had a great day, and after wearing out the hospitality of several other large aeroplane manufacturers, including Bristol Aircraft, we took off around London on a pub crawl, getting totally lost. I'm generally a pretty good navigator, but my internal compass blew a fuse trying to sort out London streets. Another thing that completely defeated me was the old-style English money — pounds, shillings and pence — I would just hold it up to the bartender and say, "Take what you want." They were all quite scrupulous and would say this is so much and so much, tuppence and bobs and all the rest of it. While we were careening around Picadilly the king and queen drove by within five feet of us and gave their little wave. We looked around and couldn't find anybody else, so it must have been intended for us. I'm sure we looked very impressive loitering about in our Woolworth's sport jackets and no overcoats.

At one point we found ourselves at St. James Palace, which was all very well, but uppermost on our minds was that after swilling all this ale we both desperately needed to take a leak. So Hugh went up to one of these guards in the big hats standing outside the palace gate and said, "Say, Mac, can you tell me where a guy can take a leak around here?" Well, these are the guards that walk around like robots not looking to the left or right and you're not supposed to talk to them, but we didn't know this. This guy just kept looking straight ahead. So Hugh started going through his vocabulary for all the other ways of describing what we wanted. Lavatory, biffy, john, pissoir — finally the guard opens the corner of his mouth a crack and says, "Just pee down the light well, guv'nor. We all do."

We made our way back to the Dorchester about two o'clock in the morning. We were on the fourth floor, and from the lift the hall seemed to stretch out about a quarter of a mile to our room. And all the way along it, outside each door, were a pair of men's shoes and a pair of ladies' shoes, a pair of men's shoes and a pair of ladies' shoes. Hugh stopped and looked.

"What did that goddamn book Bud give us say about this? These people must be a bunch of hicks!" So he picked up a pair and he went and put them down at another door, took that pair, put them down at another door, and we spent a very happy fifteen minutes switching everybody's shoes. Then we went back to our room and slept the sleep of the innocent.

The next day I caught the train out of St. Pancras Station for Derby to see my aunt and various relatives for the first time since I'd left forty-four years earlier. They demanded to be brought up to date on all my latest accomplishments in the world of commerce and appeared to approve the way I'd carried out directives made at my nativity by my long-lamented grandfather all those years ago. I wrote Dad on November 12, 1950:

> I find Aunt Bella exceedingly well. Still very active — never sits still. She is very fond of cathedrals. If she had her way we would spend about a month just visiting cathedrals all over England. Personally of course I am not too interested and feel the country is a bit cathedral-infested. Every direction you look — steeples steeples and more steeples. And the people haven't got decent houses to live in! Generally speaking I find England almost exactly as I expected — in fact so exactly as described that I feel I have known it all along — little trains, hedges, thatched roofs, and more churches than Quebec! The houses are so cold you have to go outside to get warm occasionally. It's the first time I've slept with a hot water bottle and appreciated it. The food, what there is of it, is terrible — rabbit, pheasant and fish. Coffee is straight poison, but the bread is good. The only trouble is that unless you get there first they insist on toasting it so hard it is like eating fibreglass.

After freezing to death for three days I joined Hugh Mann back in London and on November 16 we caught a British European Airlines (BEA) flight on a DC-3 for Norway. The plane was remarkably shabby, but what astonished us more was to discover the pilot circling around over Oslo quite obviously flying seat-of-the-pants style and looking for a hole to get down under the cloud layer. We didn't allow our pilots to do that even in single-engine floatplanes! In Oslo we were met at the airport and taken to the Grand Hotel, where we were greeted by General Riiser Larsen, president of SAS. We were terribly honoured. We had all heard a lot about him when he was in charge of the pilot training program in Canada during the war. They had trained hundreds and thousands of pilots—it was one of Canada's big contributions to the war.

The general made a great fuss over us and immediately fetched us up tumblers full of akvavit. Hugh and I sipped at the stuff and thought we'd got into turpentine, but the general stopped us. "That's not how you do it," he said. "Watch me." He stuck his chin up in the air, opened his mouth, there was a sound of escaping air or something, and he dumped the whole glassful, ka-chunk, right into his stomach. He wiped a tear from his eye and said, "*That* is the way you drink akvavit." We felt obligated to follow his example, but it damn near killed us.

The next day they made all the arrangements for us to see the Sandringhams. They had been operating these big British-built four-engine flying boats very successfully all the way up the coast of Norway right into the Arctic. They were a commercial conversion of the wartime Sunderland flying boat and they were huge. I could be mistaken, but I think they carried fifty or sixty people. They had got them to the point where they were operating on instruments. They were way ahead of us in instrumentation. They'd come into a fjord and do a letdown through cloud, homing in on the Omni Range. Omni Range became common later, but this was the first time I'd seen it close up. You have two radio beams swinging around a land station, one going twice as fast as the other. When you pick it up you can tell your position by measuring the difference between the two beams. This allowed them to squeeze these big boats in between the narrow coastal crags quite safely in bad weather. They also used DME. Distance Measuring Equipment. It is a signal-interrogation system. You have a little transceiver sitting at the station. You have a transmitter on the plane. It transmits a pulse to the station and the station transmits back. The time differential gives you your distance to the station. Then they used radio altimeters. These bounced a pulse off the water, which gave them their *measured* height. Barometric pressure wasn't accurate enough. They had their exact distance to the water, their distance off the station and their radial going in. There was fifteen to eighteen hundred pounds of radio equipment stowed aboard each plane. They were capable of landing absolutely blind with this, on the water. The accident record was excellent. It was amazing. I don't know of any other instrument operation in the world that wasn't on land.

The only thing it couldn't tell them was if there were any

obstructions. In connection with this they had to have very good shore control and patrol the area to make sure there was no drift or small boats in the landing zone. When the big ships came into the dock, they were pushed in tail first by two small boats. We were very interested in this whole process — not that we could have used it on this coast, the DoT would never have gone for it. The Sandringhams appealed to us very much, however. We thought, Vancouver to Prince Rupert once a day — we can fill these damn things.

They had a whole fleet of them. The all-up weight of the aircraft was fifty-six thousand pounds, full load takeoff in calm water fifty-seven seconds. They climbed at four hundred feet per minute. They got about a thousand hours per year in aircraft utilization. They beached them every 150 hours, changed engines every 750 hours, and spent 80 hours maintenance per flying hour. The service depot at Stavanger employed a hundred men. They operated normally from April 1 to October 1. Then they closed shop completely for six months while the whole country was frozen up. It was to lengthen the season and not for any other reason they were going to the effort of constructing airstrips all up the coast and switching to DC-3 equipment. On the coast of BC, the Sandringhams would have all the same advantages, dodging up and down inlets where there were no airstrips, and run year-round, like a steamboat with wings. We were sold. We could barely restrain ourselves from placing a couple on order, but we held off until we could get back home and do some checking, and it was a good thing.

The hangup was that these big buzzards drew seven feet of water and there wasn't clearance for them in the Middle Arm of the Fraser where the Vancouver seaplane base was located. We tackled the DoT to see if there was any possibility of dredging a channel for them, pointing out that this might save millions in airport construction up the coast, but all we got from Ottawa was yawns. That was the end of it, and the end of our attempts to go on building the airline around the flying boat concept. From this point on, we would be moving to amphibians and then to land planes, just as the Norwegians were doing themselves. But the memory of General Larsen and his sharp IFR operation over there on that other twisty, foggy piece of coastline would remain with me as an example of what we, too, might someday achieve.

ELEVEN

The Alcan Project

I GUESS IT WAS IN ABOUT 1948 we first began to hear about the big Alcan aluminum project. The Aluminum Company of Canada had begun as a subsidiary of the US-based Aluminium Company of America but went on its own in 1928 and during the war became one of Canada's largest corporations. Its original plant was located in Shawinigan Falls, Quebec, and in the forties massive new works were developed at Arvida. After riding out a post-war slump in metal demand, Alcan was parlaying new Cold War defence concerns into another big expansion. It began planning a new smelter near a source of cheap hydroelectric power. Indonesia and Australia were mentioned as possible sites. Federal Trade and Commerce Minister C.D. Howe, said to have more power than Prime Minister St. Laurent himself, had particularly close ties with Alcan through its president, Ray Edwin (Rip) Powell. They were both from the same US backgrounds and had attended American schools together. When Howe's dictatorial style finally got both himself and the Liberal Party thrown out of office in 1957, he stepped into a directorship on Alcan's board. And while Howe was still in Ottawa, Alcan always seemed to have the inside track with the government. The huge Alcan expansion of the war years was financed in part by special federal tax concessions allowing accelerated depreciation to the tune of $179 million. All wartime industry was eligible for these same concessions, but Alcan got over one-third of the amount allotted for the entire country.

In 1949–51, Howe appeared in the guise of a patriotic crusader trying to keep the new Alcan expansion in Canada and teamed up with the government of BC to promote smelter sites along the west coast. Throughout 1949 and 1950 there was much excited speculation in the media about which exact location they might choose. Bute Inlet seemed to get the most attention early on, but that was because the real frontrunner was in an area so remote and little known nobody even knew how to describe where it was. Kitimat was just a tiny Indian reserve at the head of Douglas Channel, one of the coast's less frequented inlets, four hundred or so miles north of Vancouver. And only a seineboat skipper knew where to look on the map for Kemano Bay. These were the two neighbouring sites finally chosen for the new smelter and powerhouse. In order to seal the deal, Howe and BC Mines and Resources Minister E.T. Kenney signed away a fifty-five-hundred-square-mile watershed in the Nechako Valley region and arranged another $170 million in accelerated depreciation. In BC, the cabinet rammed through a special provincial act called the Industrial Development Act, which made Alcan supreme masters of the territory, with power to short-circuit normal land taxes, stumpage fees, water licence fees, mineral rights, native land claims and other resource management regulations. Before the end of 1950, a $160-million megaproject was announced. There would be dams, a ten-mile penstock bored through a mountain, an A-bomb-proof underground power station, a fifty-mile transmission line, a full-scale smelter and an instant town situated on the Kitimat River delta.

At the beginning of 1951 Alcan signed a deal with the American construction giant, Morrison-Knudsen of Boise, Idaho, to build the power plant and tunnel. Mannix Ltd. of Calgary, an M-K subsidiary, was signed to build the main dam and a consortium of BC contractors under the name of Kitimat Constructors was given the job of building the townsite. I'd forgotten most of these details, but there is a very good recent writeup on all the politics behind the Alcan project called "The Sovereign State of Alcan" by Richard Overstall, in the *Telkwa Foundation Newsletter* Vol 6, No. 1.

It looked like one hell of a flying job to us. The whole area was landlocked — there were no roads and no steamer docks. In 1947 we had been granted a Class 2 non-scheduled licence, giving us Kitimat Indian village as a whistle stop on our regular service coming down from Prince Rupert to Ocean Falls, Alert Bay and on in. Class 2 was like a charter in that you had no fixed schedule but you were entitled to charge on a per-seat basis or a per-pound basis. It had advantages, particularly to the customer. You could take a six-place plane out, pick up three passengers and charge them so many dollars a passenger — according to your filed tariff. You didn't have to charge them for the whole plane at once as in straight charter. But you were not obliged to make regular trips on schedule. With Class 1 you were really tied down tight and had to subscribe to all sorts of rules and regulations. We were operating under all four classes. With Class 4 you could only charter the whole aircraft and go when you want where you want, as long as you didn't interfere with anyone's protected routes. Class 3 was somewhere between — you could charge on a per-seat basis but the demands for regularity as well as the protection from competition were less. You were also restricted by size of plane — on both Class 4 and Class 3 you were normally restricted to Group B and C aircraft, meaning you couldn't fly larger planes like Stranraers. With Class 2, you could. So Class 2 was the ideal licence from our point of view, and that's what we had over the whole Kitimat-Kemano area.

As far as we could see, we had the Alcan project sewed up. The construction companies had no choice but to come to us for all their flying needs. We were going to be busy beyond our wildest dreams. This was going to be real money, like nothing the coast had ever seen. This would be the payoff we'd been waiting for. Already in 1948 we started to get some of it. We sent Norseman CF-EJB into Whitesail Lake under a pilot named Hughie Hughes and he stayed the rest of the season flying for Northern Construction and BC Engineering, doing preliminary survey work. We billed them fifteen thousand dollars and they gave us a letter saying how much they appreciated dealing with us. This was the initial flying done on the Alcan project.

In 1951 we decided the time had come to make the big leap to Canso aircraft, which we had been talking about doing for years. The PBY or Canso was the aeroplane which replaced the Stranraer on coastal military patrol and was a much more modern aircraft, although it hauled the same number of passengers. But it was a monoplane, it flew faster and it was amphibious, meaning we could take off on land at Vancouver, Comox, Port Hardy and Sandspit while landing on the water at Kitimat and Kemano. Some models could also be certified for IFR, or instrument flight rules (IFR), which would allow us to fly when it was foggy or dark and give much more dependable service during the bad weather, providing we could get the government to install en route navigational beacons. We looked around and found a Canso being offered by the Hudson's Bay Company Fur Trade Department in Winnipeg. It looked decent and they were willing to let us have it for fifteen thousand down on a total purchase price of fifty-five thousand, so in April of 1951 we bought it.

In the course of negotiating the sale we came to know the assistant manager of their transport division, a guy named Duncan McLaren. I was impressed with this man. He was a pilot, an engineer, an administrator, and he was sharp. He looked sharp. He had slicked-down hair like George Raft and piercing eyes that seemed to know what to look for. Somewhere along the line he'd picked up a quality I was slowly coming to recognize as *airline thinking.* As opposed to bushpilot thinking, which was what we mostly had. You could tell just from the kind of words he used — talking about spine-and-feeder systems, equipment trust financing, common rate points, over-all ton-mile cost — that he was one of the new breed. How he picked up this outlook working in the fur trade division of the Hudson's Bay I was never sure, but I was quite convinced it was just the thing we were in greatest need of at that particular stage of our development, so I made him an offer. He leapt at it. I didn't know it then, but he'd been involved in some sort of power struggle at HBC and he'd been looking for a new home anyway. I had in mind that he would be just the guy to put our operations manual in shape and I was right. I installed him as my executive assistant and he went right to work.

In 1949 the Prince George bush flier Russ Baker had got involved doing some flying for Alcan from the Prince George end, which was natural because the Nechako-Bulkley Valley area was really his bailiwick and he knew it better than anybody. He put another pilot or two on the interior work and applied for a Kemano charter base down at the coast for a maximum of six small Group C floatplanes. This was our territory, and we vehemently opposed his application. I'm not sure now how it went, but I think Bud Lando got the message from Alcan they would consider it a favour if we would agree to let this small operator who was proving useful to them have an ice-free base for their winter operation. Bud came to me and said, "Look, the Air Transport Board can't allow Baker to set up a coastal base in our territory as long as we object, but we're really placing everybody in an embarrassing position by playing dog-in-the-manger here and it could cost us goodwill in the long run. I don't see what we have to lose by letting Baker have this base as long as it is very clear he can use it only to carry out interior work." Some of the rest of us had our doubts, but we had to admit this was mostly because of our personal dislike of Baker himself, not because we saw anything terribly wrong with Bud's logic. We let Bud talk us into dropping our opposition, providing Baker gave both QCA and the Air Transport Board a formal undertaking he would never venture down the coast, which he gave readily, and the Air Transport Board awarded Central BC Airways an intermediate charter base at Kemano.

Bud said wonderful, this will cement our good relations with Alcan and their major contractors Morrison-Knudsen and Kitimat Constructors. Everyone was congratulating us on this. The ATB was listening to us complain that we were getting shafted on mail pay and they were replying, "Oh well, we're not worried about you fellows anymore because we know you've got this bonanza coming up at Kitimat."

The only thing that worried us was a rumour I picked up from both Alberni MLA Jack Gibson and ATB chairman John Baldwin to the effect that Alcan was thinking of starting its own internal flying service. When I was visiting Ottawa in July, 1951 Baldwin urged me to detour through Montreal to visit Alcan headquarters. It's something I would have just as soon have avoided but I decided it might be worth it just to bolster our position.

Alcan's offices were located on the eighteenth floor of the Sun Life building. Some of the top brass turned out to meet me: McNeely Du Bose, vice-president in charge of BC operations, A.W. Whittaker, general manager in Montreal, P.E. Radley, assistant manager in charge of purchasing and traffic, and Whittaker's son Dick, who was to be in charge of BC air transport. I had a lengthy talk with Radley, gave him the background on QCA and reviewed the work we'd already done for the Alcan project. Then I outlined all the ways we wanted to improve service for them. I explained we'd bought a Canso exclusively for their work and were arranging to buy more as needed, the idea being they could take off Vancouver on wheels, alternate Port Hardy on wheels, landing at Kitimat on the water. We had already made representations to the DoT for a radio beacon at the mouth of Douglas Channel so Cansos could fly to within fifty miles of the construction site on instruments. This would mean much more reliable service with fewer cancellations due to weather.

Radley cautioned there would still be a need for small floatplanes running from Kitimat to the surrounding area. I said that would be no problem since Kitimat was a licensed base of ours and we were ready to station aircraft and maintenance crews there as soon as they were needed. I asked him outright about the rumoured company air service and he dismissed it out of hand as completely groundless. I tried to feel him out as to whether we were on the right track, and both he and the others gave me to understand QCA was everything they were looking for. Radley said they had been doing some business with Baker, but his equipment was too old and small to be very satisfactory. They urged me to work closely with Morrison-Knudsen, their principal contractor, saying that M-K's needs had to be given priority in all instances and that I should take my direction from the contractor's dispatcher in Vancouver. I assured them if M-K only let us know what they wanted, we would knock ourselves out to oblige. I went away very reassured, feeling I'd done a pretty good public relations job.

So it was a total bombshell when, later that same month, Baker announced he'd signed a private contract with Morrison-Knudsen to provide flying services to them for the next two years. I'd received no hint of this in Montreal.

Now, one area where Baker had it all over us was politicking. He personally flew the Liberal-Conservative coalition minister of mines and resources, E.T. Kenney, all around the interior during the 1949 election. When the coalition got kicked out in 1951, Baker switched his politics to Social Credit and put himself at the service of the new minister of highways, P.A. Gaglardi, who availed himself to such an extent he was known ever after as "Flying Phil." When I heard about things like this, I thought it was just dreadful. It never occurred to me that anyone whose support was worth having could be bought with such cheap stunts, but that's where Baker's approach and mine were at opposite poles. From the start in the radio service business I figured if you buckled down, worked long hours, treated everybody honestly and delivered a quality product, success would find you. This was a hangover from my dear old hard-working dad — who, of course, never made a cent. I should have known. We had problems breaking into big-time aviation with no capital, but we were doing our best, and the idea that kept me going was that some day we would get all good planes and all good pilots and first-class procedures and everything else would fall into place. Coming from a technical background, my emphasis was all on the operational stuff, the delivery-of-service side.

Baker took the opposite approach. He didn't care what kind of service he delivered as long as he was able to keep favour with people at the top. He didn't give a damn what kind of job he did flying forest fire patrols as long as the people who handed out the money kept thinking he was a great guy. The sort of thing he reserved his special efforts for was taking H.R. MacMillan out on hunting trips and sending gifts to C.D. Howe. He was one of these people who loved to drop names of all the important people he knew, and from the earliest days he saw the potential of flying as a way to meet the rich and famous; he played this for all it was worth.

I was brought up in the old English style which taught you that if you came from good family and behaved like a gentleman, that was enough; you could count on the right sort of people to see it and you didn't have to sell yourself. This was reinforced during my stint working with BC coast loggers, who would poke a bigshot in the nose before they'd take any of his guff, so I tended to look down on people who went chasing after VIPs, but Baker wasn't as dumb as I thought. What he was doing was building up a network of people with power he could call upon when needed. It was the political approach. He was a shameless egotist, but that wasn't the only reason he worked at keeping a high profile in the media; he understood much better than I how the media could be used to create political heat. He kept a suite at the Ritz Hotel with a well-stocked bar and this is where he entertained. He even invited me up there once. "Come on over," he said. "Join us for some fun. We've got some girls you might be interested in."

The thought terrified me. My idea of entertainment was to have ATB chairman John Baldwin over for a home-cooked dinner, after which we'd go down to the basement and play with my model train sets. We didn't spend money on people like Baker did. And when he smelled a Kitimat contract in the wind he pulled out all stops and campaigned until he got what he was after. My dignified interviews with head office officials just weren't in the same league as the three-ring circus he was running. When Baker won the contract he came around the airport crowing. "You guys are out!" he said. "You might as well send that Canso back to Winnipeg. I got the whole thing sewed up right here." He patted his breast pocket. Oh, he was obnoxious.

We were thoroughly alarmed by this turn of events, which caught us right off balance. But we really couldn't believe they were serious. The bulk of the business was obviously going to be between Kemano and Vancouver. We understood Baker couldn't fly between them, not only because of his solemn undertaking to us, but because it wasn't permitted under his licence. But more than that, it didn't make sense to even attempt it. For one, Baker was a seat-of-the-pants guy still running practically a one-man operation. He didn't have the equipment or the personnel to handle a job that promised to be one of the biggest industrial airlifts in Canadian history up to that time. And why bother with him? The logging companies had to subsidize our Charlottes operation in 1945 because there was nothing else available. Alcan, on the other hand, had a competent, experienced, well-equipped operator already in the field—QCA. Why charter a whole plane from Baker at a cost of five hundred dollars to fly one little injector or gauge when they could just as quickly send a five-dollar parcel on our regular flight? It didn't make sense. We felt it would never work out, and time would soon prove that. In any case, we still had the only licence serving the area from Vancouver and would obviously get most of the business on our sched flights, which we would beef up.

Dunc McLaren was quoted in John Condit's *Wings over the West* as saying "Spilsbury had the old pioneering spirit, I guess, in putting it together, but once he got it semi-glued, he didn't know what to do with it."

I will admit there's some truth in this. In one way I knew exactly what to do with QCA: I wanted to make it the best damned airline in North America. I knew that our Stranraers, running on the old CPAL sched routes plus all the routes we'd added, had by this time darn near put the old Union Steamship Company on the rocks, destroying the coast's old lifeline. I felt burdened with a deep responsibility over this, deeper than maybe anybody else would have, because I was one guy who knew how essential those steamships had been for coast settlers like my parents. It was my own intimate knowledge of the coast that had pushed our air services into nooks and crannies like Sullivan Bay and Minstrel Island and Seymour Inlet, and made it such deadly competition for the steamer service. Another person might not have bothered or not been able to make a purely coastal flying service work; Grant McConachie, for instance, had never been interested in it and had passed right on over it on his way to Prince Rupert, Whitehorse and the Orient. But I had taken the coast on and having done that I wanted to put in place an air transportation system that would be the best of its kind in the world.

Norseman upset by snow build-up on wings, Vancouver airport. Mike de Blicquy pokes holes in fuselage with pike pole as it rises.

Anson piloted by Bobby Duncan overshoots runway, Somass Airport, Alberni.

Johnny's favourite Norseman, CF-CRS, dunked with seven passengers by pilot Cedric Mah.

Canso nosed over in ditch during maintenance move at Vancouver airport.

That's what I wanted, but McLaren was right enough in saying I didn't know exactly how to bring it about, particularly on a shoestring budget. Neither did he. He had some ideas, but it is safe to say nobody in the world had been confronted with quite the same problem we had in trying to build a first-rate Class One airline in a place like the coast of BC, unless it was SAS flying IFR up the fjords of Norway.

I knew we had a long way to go. One of the many things people thought up to go with our initials was the phrase "Quite Certain to Arrive." My old friend Arthur Lipman treated me to an ode on my eightieth birthday in which he recalled QCA with the lines, "You've run an airline on a dime/ Which never once arrived on time!!/ Just merely mention QCA/ And brave men shiver to this day..." It wasn't that bad, really. But the thing we were trying to do was nearly impossible. As Jack Miles told me — Jack is an old QCA pilot who flew on into the jet age and only in the 1980s retired as a vice president of PWA — the BC coast really isn't such a bad place to fly IFR, but as far as flying regular schedule VFR, it's just crazy to even try it.

But that's what we were trying. VFR or contact flying requires the pilot keep visual contact with the ground at all times. But the coast of BC is constructed like a huge maze. You never know what's going to show up around the next point. You're always flying up one of these narrow inlets, and during ten months of the year you're flying under limited visibility. Johnny Hatch said the first eight months he flew for us he made weekly trips up Seymour Inlet without ever seeing the mountaintops. When he finally got up there on a clear day he thought he must have turned up the wrong channel. With the top showing he couldn't recognize it. The trouble is, flying up an inlet like that, if you come around a bend and find the ceiling has closed right down on you, what do you do? The damn channel isn't wide enough to make a turn safely in a lot of places. You're lucky to have room to put the floats down on the water. Our pilots spent hours "motorboating" — taxiing along the water "on the step" after they'd been forced down by low ceilings. It wasn't a bad way to travel if the sea was relatively calm. You could get up almost to takeoff speed and make about half the time you would flying (but use up five times the fuel). If you saw a log loom up out of the murk you could just push ahead on the throttle and fly over it. If you saw a bluff, you could pull back on the throttle and drop down in the water for a fast stop.

But this kind of flying was pushing the margin of safety pretty far. On the west coast of Vancouver Island you had ocean swells to contend with. We'd send our pilots out to practise putting down and taking off in the big rollers—it wasn't quite as impossible as it looked but you still had to know the technique. In August of 1951 we had a whole Norseman, CF-GRQ, disappear between Muchalat Arm and Tofino with seven people on it. Never a trace was found. The thing was, as long as you were operating on a charter basis you could listen to the weather report, look at the sky and decide whether you wanted to go up or not. But once you had scheduled service, you had to fly the approved route or else fill out a report to the DoT explaining why. You were obliged to send the plane out there every day, good weather and bad, to satisfy the schedule and pick up the customers who were standing on floats in the rain waiting, having travelled all day in a small boat to get there. You often ended up flying when you shouldn't have been. And you ended up taking a lot of trips you couldn't complete on time, if at all.

It scared me. I wondered how long we could go on before something major happened. It was bad enough losing seven people in a Norseman, but now with our Cansos we were flying around twenty-three at a time, under all kinds of conditions. It terrified me. But I was on the outside. Johnny Hatch was operations manager and he kept me fenced out of the operations side as much as he could. It was dangerous to tell me too much, was his attitude. I could see his point, because I didn't know flying like the pilots, especially the senior men like him and Art Barran and Rupert. I'd ask dumb questions and make them feel embarrassed in front of their drinking buddies. I didn't share their background—bush flying in the north, rescuing people who were dying, night bombing missions over Milan.

Pilots were born, not made. I was acutely aware of this and it made me diffident in my approach. This was a handicap I shouldn't have had, because, really, this time, I knew as much about flying as most of them, and a lot more about running an airline. But I was like Rodney Dangerfield, I couldn't get no respect. Because I wasn't one of them. So I had to fight to have much influence on operations, and I only attempted it when I saw something specific happening that I could prove was dangerous or bad business or against the rules we'd agreed on. When I brought up vaguer problems like do our flight crews have enough of a safety-first attitude, or is our pilot discipline strong enough, do people fly by the book enough or are they still trying too many bushpilot-style heroics, I would be told to quit worrying and go back to my office and rustle up some more money so we could get better planes.

This argument dated back to 1948, following our second fatal crash, when pilot Walter Britland flew a leased Bellanca Skyrocket into a cliff at Growler Cove, killing himself and a fisheries inspector from Alert Bay named Kenneth Weaves. It was one of those company debates that stirred such deep feelings people in adjacent offices got writing each other long letters, and I led off with one to Johnny:

> Dear John:
>
> At the risk of sounding cold blooded, I must point out some vital statistics. Our operation has killed nine people in the last twenty-four months. Two fatal accidents were involved, both being bush-type operations. Overwhelming evidence points to the fact that the cause of both was pilot error.
>
> I personally feel there has been a dangerous tendency on our part to avoid adequate criticism of the pilots involved, feeling possibly that such criticism is unfair in the absence of the man to defend his actions. There has also been, I feel, a tendency on our part to erroneously regard the two incidents referred to as accidents—isolated mishaps over which we have no control. This, I am afraid, is not so. We have sufficient evidence now to indicate a definite trend in our operation, pointing up the fact that the element of human error among our pilots is much higher than it should be.
>
> My own opinion is that the pilot material in our own organization has been well chosen. The only apparent weakness would seem to be that affecting the pilot's general attitude of responsibility towards life and towards the lives of his passengers, his company and his profession.

Johnny came back the next day with a detailed letter listing all the ways in which pilots' working conditions were less than ideal from a safety standpoint. Which was fair enough. I was asking a lot of these guys. I was asking them first of all to fly regular schedules over some of the most irregular country in the world, relying totally on visual flight procedures in places that weren't even visible for months at a stretch. And I was asking them to do this in a queer collection of twenty or more aircraft, no two of which were the same and no one of which was anywhere near up to date, and some of which shouldn't have been flying. So I was in a poor position to complain.

Still, I wasn't satisfied. Johnny could have made it easier for me. I felt I couldn't really raise the kind of questions I wanted to without questioning his ability, and I just couldn't do that. I had the greatest respect for his competence, as everyone did. Johnny was confidence-inspiring. He had been decorated during the war and he never talked about it, but it fit. He was just born to take responsibility. He wasn't smartass like Art Barran or yappy like Bill Peters. He was quiet and authoritative. And he could handle people. Nobody resented his authority. He never got into jams he couldn't get out of. He just seemed to ride up over every new challenge like he'd seen it somewhere before, and yet I knew he was learning as he went along just like all of us. He seemed to have no limitations.

And maybe he didn't. The limitation was there, but it was more a limitation of his world than him personally. He typified the old style of flyer who invented the modern aviation industry but couldn't accept what it was they'd invented. Johnny's type of guy was imaginative, resourceful, courageous—heroic even. The kind of man who made bush flying famous. And more than that, Johnny was a human being. He treated his pilots as equals. He wouldn't let me turn them into rulebook-slaves. If I lodged a beef because I thought some cowboy had played fast and loose with passengers' lives, Johnny would say, "Look here, this guy is a pilot and he did what he thought was best at the time. You don't know what it's like up there." I could count on Johnny to make the supreme effort on the company's behalf at all times, but I could also count on him siding with the pilots in most cases where our interests diverged.

It was a pilot's airline while he ran it. And given that few of our pilots had previous airline experience, mostly just Air Force training, it was an individualist's airline. Our planes were certainly all individuals and all our pilots were individuals too. They'd take off and from there it was just between them and God. Well, maybe that suited the terrain, because most of our passengers were old-style west coast individualists, too. Maybe I was, too.

But the modern aviation industry, as it was only then emerging, is

no place for individualists. Pioneering a route or flying a daring
rescue mission to some isolated outpost, you had to be a free spirit,
but that's not what we were doing any more. We were grinding back
and forth over scheduled routes. Everything was the same every day.
The only variables were the loads and the weather. We weren't
facing the challenge of strange territory, but on the other hand we
were logging a thousand times as many miles as bush flyers ever did.
The challenge was no longer to pull it all together for the
extraordinary feat, but to avoid the slow wearing down of things
until they snap. To avoid the one-in-a-thousand combination of
problems that can't be deciphered on the fly, because once you start
flying thousands of trips, the one-in-a-thousand combination will be
met, and with some regularity. As I wrote in a memo to Johnny in
August of 1951, following the loss of GRQ off the west coast:
"While I would be the first to admit the standard of safety in our
operation has greatly improved in the last eighteen months and
probably has now reached a standard that is the equivalent of, or
better than, any similar type of bush operation flying now or in the
past, I believe we have reached a period in our development where
we must enter a second phase, demanding improvement in operating
methods over and above anything that has heretofore been
recognized as satisfactory."

It takes a different type of personality to deal with this kind of
flying. It takes a person who can follow procedures, double-check
every preparation, second-guess every assumption, stay within a
wide margin of safety at all times and maintain unvarying habit over
long periods of repetitive work. While still remaining alert for the
freak event. It isn't easy. It's why airline pilots are so well paid. It's
also why the old rugged individualist of the bush flying era had such
a hard time adjusting to the modern era of airline-type flying. Many
of the most famous old pilots never made the transition. And very,
very few of the old bush flying companies made the transition from
airways to airline.

The trouble was we didn't know what we were up against. You
never do at the time. You may have an inkling, you may sense that
something is unusually difficult, and you may question the methods
that have worked in the past, especially if you're standing slightly off
to the side as I was, looking at operations, but you're basically
groping in the dark. Change has to come either through incredible
good luck or through heartbreaking disaster.

I struggled with this. I read books from the US about how airlines were supposed to be run, *Commercial Airline Transportation* by John Frederick and *Air Transportation Management* by Joseph Nicholson (I still have them) and they actually were a lot of help. I brought in outside experts, like Duncan McLaren. But we still had the problem of trying to break through to the pilots.

McLaren went right to work on our operations manual. In a scheduled airline, every procedure for flying, every route, had to be written down and every pilot had to read it and sign it before each time he took off. Every facet of the airline's work had to be covered by a rule in the manual, and even ground crew had to sign in and out of the book before starting or completing a task. For instance, here is a section from the Traffic Reference Manual regarding announcements:

1. GENERAL

Flight departure announcement will be made by the Passenger Agent when ramp control advises the flight is cleared. If flight departure is delayed, ramp control will advise the reason for and duration of delay. The new departure time will be announced by the passenger agent. When making the announcements over the P.A. system, it is imperative that the agent speaks slowly and distinctly.

METHOD OF ANNOUNCING

1. Departure (land planes)

"Your attention please—announcing departure of QCA flight 15 for Nanaimo with limousine connections Nanaimo, Ladysmith and Duncan. Passengers will board through gate 4, no smoking, all aboard please."

2. Departure (sea planes)

"Your attention please—announcing departure of QCA flight 31 for the west coast of Vancouver Island. Passengers for Muchalat, Nootka, Tahsis, CeePeeCee, Chamiss Bay, Zeballos, etc. (as shown on manifest) will board the limousine at the entrance of this building."

Issued: December 1, 1951
Initial when read _____

Each manual was in effect a legal document which could become evidence in an enquiry or investigation, and its owner had to sign every updated page as it was received. Each leaf had to be carefully numbered to account for any changed or deleted pages, so there could never be any doubt that it was current. If a procedure were rescinded, leaving a page missing, a blank page had to be put in its place with the inscription, "This page intentionally left blank, March 20, 1951." At least this kind of detail was what the government wanted. The gang we had up to 1951 tended to kind of laugh the whole rulebook thing off as being the obsession of smaller minds, and the Air Transport Board was after us constantly about improving our manual, but I guess we didn't exactly know what they were talking about. Our operations people didn't want to know, some of them. I knew I had to do something, the ATB was threatening to lift our licences if I didn't, especially any time we had a crackup, and that was why I hired McLaren. He was very good at writing the manual, but we still had the problem of getting the pilots to use it.

I've told about the fun side of the thing getting it started, but the real hard work came now in trying to mold this ragtag bunch of obsolete aircraft and wandering sky bums into a reliable modern airline.

The flying crowd. I didn't know what to make of them. It was a whole different world from what I'd known chugging around in the protected confines of the BC coast. Flying attracted a special group of people because it promised escape and excitement, and they lived by their own rules. We had a lot of damn good-looking women working for us, and God, they didn't last longer than a snowball. Pilots were sleeping with stewardesses, traffic staff and office girls — I didn't know just how much of this was going on at the time. Most of it I would hear in a roundabout way. "Oh yeah, that guy, he was sleeping with whats-her-name." A senior pilot was living with one of the stewardesses for a whole year and I never knew.

We had one pilot there who amazed me. He slept with a different woman every night. It was just natural for him. I travelled sometimes with this guy, and we'd go out and land in some place, say Ocean Falls, and stay overnight in the hotel. By mid-evening he'd find somebody. "I'll see you in the morning," he'd say, "I won't be using my bed." Just automatic. Thought nothing of it. He went up the coast hopping from one bed to the other. Married, single, he didn't care.

He was living in a rooming house down in the West End at one point and I had to stop and pick him up on the way to the airport. I knocked on his door. No answer. I knocked louder. Then I heard him shout, "Come in! The door's not locked." I opened the door, and here he was in a double bed with a woman taxi driver.

"I'll just be a minute," he said.

"Why don't you go into the kitchen and make us a pot of coffee?"

He wouldn't stay with them, just one-night stands as they say, but they all thought the world of him. He had that something about him. They'd come down from up the coast, having walked out on their husbands—"Here I am." One thing, he'd never boast about his conquests. The other pilots were all so damn jealous of their own reputations, it was all they could talk about. Rupert would spout off, and he and Bill Peters would get vying with each other to see who'd get the woman, but they were so clumsy usually neither of them would get anywhere. There was a lot of hard feeling among many of the pilots. Then there were others whom nothing could distract. Art Barran was the next thing to a robot in the air, I thought. He never made a slip or bent a plane on us, and if he had a sex life he never let it show.

But this fast living was all totally new to me. I'd never seen anything like it, and I don't suppose it was very typical of that period, but the airport crowd was, as I say, special. I didn't believe in it, I kept out of it myself, but it was unsettling. There were many who survived as families, but a considerable number divorced at least once or twice. Breakups were occurring all the time and this was a real worry because of the mental condition it placed key staff in, especially pilots. You get someone who's just lost his wife or girlfriend at the controls of a plane with a load of people entrusting their lives to him and how stable is he? Does he have any suicidal tendencies? We had a pilot stick an Anson into the mountain behind Halfmoon Bay and we suspected it was because he was depressed about his love life.

I engaged a company who specialized in psychological assessment to test all our pilots and top executives, including myself. I couldn't bring myself to accept all that they claimed to have uncovered, but looking over the results later, I saw they'd fingered those pilots who would eventually wind up killing themselves almost to a man. It was like divining. It was scary. Bob Gayer often talks about that. He had the same thing done on his group at Associated Air Taxi, and he can go down the casualty list, ticking them off. Every one of those guys the tests said were not safe eventually ended up in trouble. Not deliberately maybe, but because they lacked the ability to think a hundred percent flying when they were flying.

Fighting obsolete aircraft was one thing, but dealing with people was even harder for me. It was as hard as calling up my cousin Rupert, who with his vision and flying knowledge was in every way responsible for the airline existing, and saying "Rupert, you're fired." But I had to. Rupert would never drink on the job and he'd never bent a plane of ours, but he'd drink at night. He'd go into a small place like Prince Rupert or Stewart and hit the bars and bullshit and scare the hell out of people whose confidence we were struggling to build up. I had so many complaints I finally had to get him on the mat and say, "No more." I was trying to enforce the same rules that TCA had: no alcohol twenty-four hours before takeoff, but it was an uphill battle with the gang I had then.

Johnny Hatch wasn't a very good example, I'm afraid. Neither was our chief pilot, Len Fraser. He'd been let go by CPAL for drinking, but Grant McConachie, when he was trying to convince me what a first-class pilot Fraser was, claimed he'd taken the cure. Maybe he had, but it didn't stick. One time in particular he took a Canso into Garibaldi Lake and got so goddamn plastered he couldn't fly it out. Hugh Mann had to take over. I demoted Fraser and put Art Barran in his place in September of 1951.

And poor old Charlie Banting, who had plucked us from the teeth of disaster back when we were rebuilding the Waco and without whose vast engineering expertise we could never have gone on. The dearest guy you could ever want to meet. But his suspicions about the cause of his drinking problems proved all too true. Here he found himself superintendent of maintenance for a far bigger airline than before, and the responsibility took a cruel toll. He'd get so goddamned plastered for a week you wouldn't see him. He couldn't make it out to the airport. It wouldn't matter even if we had an aircraft coming up for his signature, we'd have to make excuses to the DoT, "Sorry, our chief engineer is 'indisposed.' " I can't say what it did to me to drop the axe on poor old Charlie. But it had to be done, and I was the one who had to do it.

As it turned out, I wasn't half tough enough.

TWELVE

Mount Benson

WE HADN'T GONE FAR INTO 1951 before we realized all was not well on the Alcan front. We had stretched our necks out a long way and bought a second Canso from the Babb Company for sixty thousand dollars, to just make darn good and sure we were ready for the business, but we weren't getting it. What was really happening we could hardly believe, but there was no mistaking it. Baker brought in a big fleet of leased, subcontracted, chartered and borrowed floatplanes and he was using them to break both his undertaking to us and — we maintained — violating the terms of his charter licence by flying into Vancouver. Between July and December of 1951 CBCA's swarm of gnats hauled 5,556 passengers and 239,214 pounds of freight to the project area, while our big birds lumbered up and down the coast with only 4,031 passengers and 176,000 pounds of freight. Four thousand of the CBCA passengers were out of Vancouver, according to Baker's own records.

The legality of his flying to Vancouver at all was a matter of dispute. He maintained that under private contract he could fly anywhere he wished. We maintained, in an official complaint to the Air Transport Board in Ottawa, that "this service, for all practical purposes, supplants the Class 2 service between the two points that QCA is licensed to operate." The Air Transport Board appeared simply confused, and after watching Baker fly for several months without their approval, decided the easiest way out was to sanction his contract with a Class 5 licence. ATB chairman John Baldwin would later confide to me that this was a serious mistake.

Now I realized what a blunder we'd made in ceding Baker that foothold at Kemano. Not only was he mopping up the lion's share of the Kemano traffic, he was also poaching on our scheduled business at Kitimat, where activity was just starting. On his Alcan contract Baker was paid for the round trip from Vancouver. The company was thus entitled to use the aircraft going back down as well, but much of the time at this stage of the game these craft were coming down empty. It didn't take long before Baker was picking up the passengers that he'd find waiting for our Canso on the dock at Kitimat. With him they'd get to Vancouver ahead of time in most cases, and if our fare was fifty dollars, he'd offer to take anyone for twenty-five. It was all found money for him. We discovered that we were going up with loads and coming down half empty. It was very worrisome, because our main problem as a scheduled carrier was to keep our load factor up. We needed fifty-five to sixty percent before we could make money and we just weren't getting it.

Of course, selling individual fares to Vancouver was clearly illegal for Baker, and we found ourselves in the curious position of running to our old nemesis the DoT, urging them to get their rulebooks out. We put great store in the strictness which had given us such difficulty in previous years, but we soon saw a different side of government.

I may remind you that the purpose of the DoT's civil aviation division and the Air Transport Board when it was set up during the war was to control *uncontrolled* competitive flying. This had killed too many people. They had proved that already. They had proved it right here in BC before the war, when Ginger Coote and all that bunch were scrabbling for business out of Zeballos. Each pilot would fly into tougher weather than the other and take greater risks, and it only added up to unnecessary loss of life. As federal transport minister Lionel Chevrier explained it in his speech to AITA in 1952, "Therefore the government laid down a policy which was designed to allow each operator to establish himself unhampered by the chaotic conditions which flourished during the thirties." The Air Transport Board was set up to license certain people on certain routes and tell other people, "Get your own routes but don't interfere here." Eliminate competition.

Oldtimers in the aviation business have a saying that government thinking on competition versus controlled flying reverses itself every twenty years. You start with competitive flying and carry on until public outcry over the carnage forces the government to bring in controls. But as soon as they do that, the guys who are left out start crying free enterprise, and more people join the chorus until the government tears down its system of controls and the carnage cycle starts over. As I speak, in the late eighties, we are in the flush of a new free-for-all era. It gives me the shivers.

We wrote letter after letter documenting our complaints against Baker to the Air Transport Board, but nothing happened. They gave every sign of being very concerned, but because of what this project meant to Canada and C.D. Howe and all the rest, they found it politically expedient not to move too quickly.

Once we saw what the situation was going to be, we knew we had to compete or give up; we couldn't afford to give up so we had to compete. This was when I found out what kind of spirit our gang had. They were all for driving Baker back into his hole. The degree of loyalty they felt toward QCA was really surprising, given the internal problems we always had. But esprit de corps is something that happens almost in spite of itself, and nothing helps it along like a challenge from outside. Our people had been used to thinking of themselves as working for the leader on the coast, the only company with airline pretensions after the big two. This was the position ceded us by the other established companies like BC Airlines. Now to have a rough upstart like Baker come along and throw mud in our eye — it really got our gang's blood up. As far as we were concerned Baker was an impostor, a bullshitting bush flier who didn't deserve to be mentioned in a conversation about real airlines. He had puffed himself up very temporarily and artificially by getting in with this Alcan bunch and had a lot of people fooled into thinking he was an operator on our scale, but we would soon puncture his bubble and demonstrate to the world the difference between flying and bullshitting.

It wasn't at all difficult to get our people to go all out; I would say rather it was harder to restrain them from pushing the margin of risk too far. Baker was using these small floatplanes, and our big Cansos and Strannies could land in much rougher water and handle much tougher weather. Our pilots would take advantage of this wherever they could.

Around this same time I woke up one morning and found we were certified by the mechanics' union. I had no knowledge of unions and no idea how to deal with it, but I clearly didn't have any choice in the matter. All our mechanics were now unionized. This again added to our expense. It cost us more for overtime, we had to start doing this that and the other, holding grievance hearings which were surprisingly costly and took up a lot of time. It also affected company morale because now we had union organizers in the place. The mechanics' union was a pretty good outfit however, very understanding, and I learned to live with them. The next thing that happened, we were certified by the pilots' union, Canadian Airline Pilots' Association (CALPA).

So now I had to fight a war on several fronts. One to get traffic, another to fight Baker, two unions, plus the bank—trying to get money to buy aeroplanes. Oh, gee. I can see now I was making a big mistake in trying to deal with all this on my own. It was all so new to me. It was so far out of what I'd been used to. And I was carrying too much of a load.

Hepburn had been a great help to me in the early days of the airline but had gradually wound down over the years. He had never really accepted the addition of Jack Tindall to the radio company and I'd say from that time forward became increasingly trouble-some. He gave poor Jack a bad time, and Jack was such a decent guy, it hurt me to see it. Hep continued to sit on the QCA board of directors until early 1951 and he was useful on communications matters, but only occasionally in other areas. Then the Spilsbury and Hepburn auditors came to me and said, "There's a lot of cash missing." Over six thousand dollars that they had been able to track. In those days that was several months' profit for the whole radio company. They had an investigation and pinned it on Hepburn. All cash coming in by mail he'd been putting in his pocket and not reporting. We got him dead to rights. We suspended him and stored all of the incriminating evidence in the office and changed all the locks to keep him away from it. But then a funny thing happened. One morning all the papers disappeared. Our whole case vanished.

Lando went down and saved the day, he got Hep to admit to the whole thing. He handled it very well. Hep said well, he was short and needed the money. We were still only getting radio repairmen's wages. Much less than our own pilots. Lando took this all in and said, "Just as a point of interest, how did you get in to take all the papers?" Hep quite proudly explained how he came in to talk the thing over with Jack Tindall and as he stood with his back to the door he pushed some soft wax into the keyhole. Then he went away, filed himself up a key, and that night he came down and took all the evidence. Hep was still a very capable character when he got inspired.

Lando said, "Do you realize you could go to jail?" Hep shrugged and said yeah, he guessed so. But once again, Lando the humanist shone through. "Look here," he said, "there's no point, nobody wins if you do that. Why don't you leave Jack Tindall in charge of the radio company and go out to the airline?" So, effective May, 1951, Hep was appointed communications engineer for QCA with responsibility for engineering, direction and control of all company communications facilities "whether located at Vancouver or up the coast." He had no management responsibilities but it was still a pretty big job and he actually handled it very well for a number of years. Of course he still had a fair swack of shares in the airline, so he was a bit of an anomaly, a founder and part-owner with the responsibilities of an ordinary employee. This was the period I referred to earlier, when he drifted around the QCA offices reading people's letters upside down. He took such an interest in things he was hard to keep down, really.

Even before all of this brouhaha about Hep and the missing funds, we decided the time had come to apply some scientific thinking to our system of management. In 1950, after a period of rapid expansion—more employees, more routes, more earnings, more expenses, more confusion—we really didn't know whether we were coming or going. Norm Landahl as office manager was overwhelmed and far over his head. Then we got a phone call from the bank. It didn't sound good. I got Lando in, we wrestled with mounds of paper, added everything up three times, and the answer was the same. We were out of money. Bud phoned the bank, Norm phoned the creditors and disaster was avoided, but only just. After this sobering experience we established a weekly financial meeting in which Johnny Hatch took part and was a great help. Aside from being a good pilot and a terrific business promoter, he had some past experience in accounting. We tried through this financial committee to keep a lid on things, but before long we were expanding again anyway. We just couldn't help it.

Then one day I had a visit from two gentlemen with embossed business cards indicating they were from the George S. May Company, Management Consultants. I was impressed. I called in Johnny Hatch. He was impressed. These people knew everything. It took them all of fifteen minutes to identify our basic management problems: lack of capital, missed growth opportunities, employee inefficiency—it was like they could read our minds. They went on to explain how the George S. May Company could help us. First of all they would do a complete analysis of our present business. All we had to do was give each one a desk to work at and in due course they would have all the facts and figures spread out in front of us. The charge was something on the order of twelve hundred dollars a day but it should not take long and we'd be on our way to growth and profits. But we must act now, as they were just between contracts and might not be available again for some time.

John was sold and so was I. We signed up.

The first thing that happened was these two smart characters who had impressed the hell out of us disappeared, their places being taken by a team of "analysts" who did nothing but go around counting things. One day the boss analyst asked to see me. He had our ledger. He said he was shocked to discover that I, as president, was receiving only $165 a month. Before we went any further, he said, this had to be addressed. I should be earning at least a thousand a month. I couldn't hope to retain the respect of the employees, not to mention business connections, until I received a respectable salary. I asked Johnny what he thought of this idea and it turned out they'd given him a similar pitch. We decided the plan had its appeal, but we didn't do anything until the next board of directors meeting, which was in a few days. We hadn't bothered to tell Lando about putting the company under analysis, it had all happened so fast, but at the next board meeting we introduced him to our new friends and told him about their wonderful idea. To say the least, we were disappointed with his reaction. I will spare the details of the discussion, except to say that the next day the analysts were on their way back to where they came from and my salary remained at $165.

At the same time Bud agreed that QCA's financial picture was getting too complex for amateurs to muddle their way through and it was time we got professional help. He got in touch with a company he'd heard of in New Westminster, BC Management Engineers. Two gentlemen came out to see us. One, a Mr. Chester, was a chartered accountant and very experienced; the other, a Mr. Cranston, more of a sales type and a little overweight. They looked at our setup. They recognized some problems. They didn't dazzle us like the first consultants, but the fee they quoted was a lot less. In fact it was so low we could scarcely believe they could be very good. Still, there seemed nothing to lose so, with Lando's approval, we engaged them. They worked quietly and logically through the outer layers of our financial morass, making salient points as they went.

Maybe two weeks of this and they dropped the bomb. They said we *must* change auditors. We were still using John Weeden, my father's friend from the early days on Savary Island, who had helped Hepburn and me set up our first company books in 1941. I presumed he was doing an adequate job for an old guy, but they disagreed. At the core of our financial troubles was our rather poor relationship with our bank, they said, and it would make all the difference to our bank's attitude if our financial picture were presented properly. They added that there were a number of up-and-coming auditing firms they knew of locally, and they could find us a good one. Bud Lando agreed. They came back with the name of a wide-awake, progressive outfit called Griffiths and Griffiths, headed by the now well-known communications tycoon Frank Griffiths. So we put the skids under poor old John Weeden. He was very hurt and never spoke to me again. Dad, who was now living in North Vancouver and still following my dealings with avid interest, was very upset, and I felt just as bad, but it was all in our best interests, we were told.

And so it turned out. As soon as we hired Griffiths and Griffiths the management consulting team just seemed to evaporate. G&G took over entirely and it transpired that Chester, the accountant, was a member of Griffiths' firm. He took over our account and handled it from that time forward. Was this an amazing coincidence, or a clever way of finding new clients for the accounting firm? Whichever, it was QCA that came out the winner. The firm did a very good job for us, and Frank Griffiths became a director and eventually an investor in QCA. He proved a godsend at a time when we badly needed some hard-headed business guidance and eased the burden on myself and Lando over the ensuing years. I don't know what we'd have done without him. One of the first things he did was bring in a smart young accountant named Alan Houghton to take charge of our accounting department, which Houghton promptly renamed the Treasury Department, giving himself the title Treasurer-Controller. Norm Landahl, our original accountant, moved over to become Assistant to the General Manager. Johnny Hatch, who had held that title previously, moved up to General Manager. I became Managing Director as well as President. Lando, of course, was also on the QCA board of directors and was a tremendous help in a business way, but neither he nor Griffiths knew much about flying. Johnny Hatch was my main ally in flying matters but as I said, when it came to dealing with the pilots he couldn't be as tough as he needed to be.

But at some point everything still came back to me, and there just wasn't enough of me to go around.

In September of 1951 I made a personal inspection of our Canso service on the Alcan route. We were still struggling with low payloads and I guessed most of it wasn't our fault, but I wanted to see if there was anything we could do to improve our direct appeal to the customers. I tried to think like a passenger and make notes of anything that displeased me. They were small things mostly, but I was ready to grasp at straws.

> — should serve coffee in proper cups and trays. As it is the crew just hands out flimsy paper cups which the passengers hold in their bare hands.
> — should serve sandwiches. It is a long flight.
> — there are no magazines or any reading material aboard. The Canso doesn't offer much exterior visibility so it is important to have good lighting and supply something to keep people diverted.

When I arrived at Kitimat I had a meeting with Eric Melanson, the resident engineer and boss of Kitimat Constructors. Melanson was full of complaints, which I appreciated because it at least gave me something to grapple with. He struck me as a decent sort who was genuinely concerned with helping us to improve our service, unlike most of the others up there who just gave me the runaround. He was unhappy with what he called our "coach service," by which he meant our regular schedule flight. He didn't like the mix of freight and passengers and the stops at Kemano. Passengers were frequently displaced by air freight ordered up by Morrison-Knudsen.

This was a legitimate complaint. Once just previously we'd had a Canso ready to take off and the Morrison-Knudsen agent in Vancouver ran out on the tarmac to stop us. He had a load of long lengths of pipe that had to go to Kemano on a priority basis. We explained that the plane was fully loaded but he didn't care. He insisted we delay the flight and bump passengers and baggage so these long pipes could be stowed in the aisle. The flight was held up for hours, and then the passengers had to crawl back and forth over these pipes the whole way north. One of the passengers was Alcan vice-president and project manager Percy E. Radley. He was furious — at us. It was on Alcan's own instructions that Morrison-

Knudsen was given first call on all Kitimat services but that didn't deter him. Thinking about it later I kicked myself for not making a Norseman available to take care of Radley on a VIP basis, but it just didn't occur to me. Baker would have done it in a minute, piloted the plane himself and taken along a bottle of scotch, I'm sure, but at that point I just wasn't thinking along those lines. Radley never forgave us, and we paid, oh boy, did we pay!

Melanson thought maybe Kitimat Constructors should charter direct from Vancouver to Kitimat and not try to put loads together with all other Alcan contractors, particularly Morrison-Knudsen. I tried to persuade him that we could build our scheduled Canso service up to where it would be everything they wanted, but it would be necessary to get adequate landing facilities at Kitimat. Melanson for his part was convinced that a proper beaching ramp should be built immediately at the townsite, so that the Cansos could land in the water, taxi up onto terra firma and unload, load and taxi down again. It could also be used for beaching floatplanes when weather made it unsafe to fly out. He had worked during the war building the big seaplane base at Coal Harbour, which later became the Gibson brothers' whaling station, and he knew more about the subject than I did. A man named Donald—J.A. Donald—poked his head up to squawk that a ramp was unnecessary. According to him all that was needed was to install two sixty-five-foot floats connected to the shore.

I pointed out that shore floats would be alright in relatively calm weather or southeasters but not in north winds. The only kind of float that would work for the Cansos in all weather would be a free-standing one anchored well out in the water so the ship could approach it from any side, depending on wind direction at the time, and unload passengers into a boat. If there was going to be no ramp a buoy would be needed also, so that the planes could be safely anchored out overnight. It was over a hundred miles to the nearest all-weather base, and if conditions deteriorated during a Kitimat stopover pilots were forced to fly out into it anyway or risk losing the ship on Kitimat's unsheltered beach. It was a very dangerous situation.

Mr. Donald would hear none of this. I didn't know who he was or what he did, but he spoke like he owned the joint. This was one of the problems in dealing with the entire Alcan project—you never knew who you were talking to or who was in charge of what. This man was some sort of engineer, but who he spoke for wasn't clear to me. Melanson argued that a proper seaplane ramp would be the only satisfactory solution. He said he could cut all the lumber locally and put the whole thing together for fifteen thousand dollars—peanuts to them. Donald, whoever he was and whatever his reasons were, remained strongly opposed. I later had another long talk to him and tried to swing him around. I even offered to spend some of our own scarce funds, if he could get permission from the aluminum company. Melanson took me aside and said he would make his own independent recommendation to Alcan urging the immediate construction of a ramp and apron, and suggested we do the same, stressing the safety and convenience factors.

We also discussed radio communication. I pointed out that we would have to put in our own company frequencies, our own transmitter and receiver to give our planes proper weather reporting and traffic information on the way up. This was a matter of course, even at the smallest stops along our route, as was the need for a reliable local resident to act as our agent. We had located several who were eager to do this. But in Kitimat the smallest matter was a company matter and had to be cleared through company offices. With the super-powers given them under their special government act, Alcan directly controlled everything that crawled, swam or flew for a hundred miles in every direction, and I was advised to apply in writing for everything we needed, not omitting the smallest detail.

We went along with all their suggestions and made all our requests through the proper channels, but it got us exactly nowhere. As we moved on into the fall of 1951 we still had no improvement in landing arrangements, no agent and no land-based radio to give us reliable weather information. Alcan gave no reasons for their inaction, and told us to use a radio dispatcher in the freight office, an ex-army type who was completely uninterested in the task and more than once gave our pilots faulty weather reports, nearly costing us an aircraft.

One day, out of the blue, our assistant flight operations manager Mike de Blicquy walked into my office and resigned. I was flabbergasted. I thought the world of de Blicquy and relied enormously on his cool judgment. He had learned his flying in Europe and had flown for years in Quebec and he'd seen it all. He had a very wise head, and I especially valued him as a steadying influence on the rest of the crew. If Johnny was the mainspring of our operations department, Mike was our balance-wheel. I had insisted on his being placed on the QCA board of directors.

"I'm sorry, Jim," he said, "but you're headed for trouble and I don't want to be around when it happens. I've lived through all this cutthroat business before and I know where it leads. I don't want to go through it again."

I said, "For God's sake, Mike, let's sit down and talk about this. I don't want to lose you."

He said no, his mind was made up. He was sorry, but there was nothing more he could do for me. He'd tried and they wouldn't listen to him. He was disappearing over the horizon. He took my hand and said he'd enjoyed working with me and didn't blame me for what was happening. Then he shook his head sagely and said, "Jim, my friend, you must mend your political fences. That's my parting word to you. Mend your political fences."

I brooded for days trying to figure out what exactly he was getting at. I knew the competitive flying situation with Baker was getting uncomfortable, but what could we do? Recently Baker had taken to the practice of slipping fivers to outbound workers waiting on the beach at Kitimat so they would fly CBCA. Some got their ticket paid by the company, so cutting fares didn't always mean much to them. Cash kickbacks worked better. It got to where we put Bill Watts, our traffic manager, on special assignment armed with an unlisted phone number downtown, dark glasses and a bagful of cash to fly around the Alcan project bribing passengers back to our side. Baker kept upping the ante till it reached twenty-five dollars a head. It was ridiculous, but we couldn't just let this bastard take Kitimat away from us. It was going to be too big.

I guessed Mike was worried about safety standards and some of the risks we were being forced into. I was, too, but it was hard for me sitting at my desk in Vancouver to know just how bad it was. In late summer I'd heard something that had given me quite a scare. A guy named Gordie Simpson had been working several years for us. He was a flight engineer. He was a good worker but rather opinionated. The operations management didn't like him much. He'd been flying on both Stranraers and Cansos and quite a few flights had been with a senior pilot named Doug McQueen. One day he went into the chief pilot and requested that he not be assigned to any more flights with Captain McQueen, who was then flying Cansos on the Kitimat run. Simpson said if he were ordered out with McQueen again, he would have to refuse for professional reasons. The chief pilot didn't like this, consulted with the operations manager, and they decided they couldn't allow flight engineers

Canso landing at Kemano, 1951.

Pilot Sheldon Luck with Canso load of Alcan project brass.

Canso passenger accommodations.

Canso CF-FOQ with a load of freight and Alcan workers waiting for surface transportation.

dictating how they should make up their crews so they fired him.

I met Gordie just as I was going into my office and I asked him what the devil was going on.

He came into my office and closed the door.

"It's McQueen, Jim. I've never been so scared in all my life. That guy will go in on top of heavy clouds, go down through a hole in the clouds and between the mountains, land in Kemano, and then fly up and hope he misses the mountains and gets out. He's not flying on instruments and it's just not right. I won't fly with him again, and I've told him that. He's going to kill twenty-three people one of these times and I'm not going to be one of them."

I said I was very disturbed to hear that and very sorry that he was leaving over it. I offered to call operations in and discuss it with them.

"Oh, don't bother," he said. He knew he was through and he didn't care. He just wanted me to know what the score was.

I called in our chief pilot at that time, Len Fraser. He'd been with CPAL and came to us with very good recommendations over a year before. Fraser called in Johnny Hatch and between them they tried to calm me down.

"Look, Jim, you can't be going around listening to every crackpot. They've all got their own ideas. You've got a flight department, just don't you worry about flying the aeroplanes. You worry about getting the passengers and getting the money."

The standard response. I had to believe my operations people, but brooding over Mike de Blicquy's resignation, I wondered if there wasn't something to what Gordie Simpson was talking about after all.

I was disturbed enough about Mike's remark on my political fences that I paid a special visit to the most powerful person in flying I felt I could talk to as a friend, ATB chairman John Baldwin in Ottawa. I got him aside for a private chat and unloaded. I told him how worried I was about the lack of facilities, the competition, the rapidly increasing KKK (Kitimat-Kildala-Kemano) traffic and the generally deteriorating Alcan situation. I told him about Gordie Simpson and I told him about Mike de Blicquy. You had to be worried when you lost a guy like de Blicquy. And yet on the other hand I had a damn good staff of chief pilots, operations managers, and they all told me not to worry, they were keeping everything under control, they were well aware of the pressures but they'd handle it.

But, I told him, until we got proper facilities in Kitimat and until we got this illegal pirating controlled, I was very worried. Traffic-wise we'd had a spectacular year, with record revenues in every month and some routes up by as much as three hundred percent, but we were going to burn up all our profits flying these empty Cansos out of Kitimat. With all the heavy maintenance routines and backup procedures we had to keep as a scheduled carrier, and now all the added expenses visited upon us by unionization, there was just no way we should be forced into head-to-head competition with a non-union bush flyer who did minimal maintenance, hired pilots like Cedric Mah and Jim Lougheed whom we'd fired for cause and worked them 178 hours a month, and was being allowed to ignore every safety rule in the book.

Baldwin was very sympathetic and said the board took the matter very seriously, but had to follow its own procedures. He said they'd come out and investigate and if they found what I said was true, they'd take the necessary action. In the meantime, do the best you can and carry on.

October 12, 1951 was one of those rough days I spoke of when Baker was unable to get his small planes down in Kitimat and we were able to bring our Canso in and snatch a full load. The pilot was Doug McQueen. I should mention that we had very tight radio dispatch control. Aircraft had to report in from points we'd established on our flight routes, give their position and ETA and keep renewing this. We had a really busy dispatch office — moving pins on a board to show the position and altitude of each aircraft — it was run like it should be.

I was home, it was just about dark, when I got a call from our dispatcher. He said that Flight 102, our Canso CF-FOQ, was at Port Hardy RON (Remaining Over Night). Too late getting out of Kitimat. That was fine. I wanted to be notified of anything like this. Flight 102 was supposed to head south from Kitimat about noon, and McQueen had supposedly been waiting for weather but was more likely hustling around trying to drum up a full load.

About half an hour later I got another call from dispatch.

"Flight 102 has just taken off Port Hardy, estimating Vancouver 22:15."

"Did you authorize it?" I said.

"No."

He'd simply been advised the flight was off with twenty passengers aboard. The pilot had changed his mind on his own.

I was pretty upset. I tried to get hold of Johnny and couldn't, so I called back to the dispatcher and said, "When 102 arrives Vancouver suspend the flight crew pending investigation tomorrow morning." Doug McQueen was another Air Force veteran, a stocky, jovial type in his early thirties. Operations liked him because he was very gung-ho and a good producer. He got the loads when other guys couldn't. I didn't know it at the time, but he hadn't got his full instrument flight rating for night flying. He couldn't. He was deaf in one ear. Eleven o'clock the phone rang again. The plane is in, the passengers away, the pilot suspended and a report is on Johnny Hatch's desk. This dispatcher was a crackerjack. I said, "Thank you very much."

When I got in the next morning I found the investigation had already taken place. McQueen had made the excuse that he had a medical case aboard and had been reinstated without so much as a slap on the wrist. I raised hell but Johnny Hatch said, "Look. Do you want a pilot's strike on your hands? Because this pilot is the shop steward. They figure they're doing a job for this company and they just won't go for this kind of meddling around by management."

I wasn't satisfied, but I didn't know exactly what to do. Because I knew the thing we were all aware of but weren't saying was the pilot had been trying to steal a march on Baker and he'd pulled it off. This made it all the more awkward to be tough on him.

Five days later, same flight, same pilot, same weather, same thing happens. The dispatcher calls me, FOQ has taken off late from Kitimat. The plane was filled to the twenty-three-person limit.

"Oh God!" I said. "Well, you know what to do. He's going to stay suspended this time."

But he wasn't so lucky.

The last we heard of Flight 102-17 was at 18:48. He said he was twenty miles out of Vancouver descending.

Just minutes later we received DoT reports Nanaimo residents had witnessed the crash of a large aeroplane in mountains behind the town. Several hours later the Nanaimo RCMP phoned to inform us they had found the remains of Canso CF-FOQ at the sixteen-hundred-foot level on Mount Benson, six miles behind the city. It was still burning and there appeared to be no survivors.

Sun newspaper reporter Clem Russell described the scene:

Two blazing trees high on the mountain led four separate search parties to the crash site. The plane was blown to pieces when it thundered into the sheer rock face. Fragments of the plane and bodies were scattered over 200 yards.

The tail assembly was poised on the lip of a sheer 60-foot drop. When I arrived there was nothing left of it but a charred, ribbed skeleton outlined by flames.

A red bedroom slipper, not even soiled by a grease spot, was lying beside a shattered wing.

Nearby were two pulp magazines, their slick covers glossed over by the heavy night dew.

The body of a man lay with the head on a carbon dioxide fire extinguisher. Farther away, where a wing was resting against a snag, a charred body was huddled, one logging-booted foot projecting through the undergrowth.

Another man's hand protruded from a cabin window.

Members of the search party stepped carefully around the wreckage, fearful of stepping on concealed bodies.

The glimmering flashlights of one of the party picked up an engineer's micrometer.

The three-hour hike up the mountain had almost exhausted the batteries of our torches. It was a merciful factor, for the dim lights softened the horrible scene around us.

With our traffic manager, Bill Watts, I caught a night flight to Victoria and took a taxi up to Nanaimo. We arrived at the RCMP station in Nanaimo at eleven P.M. and went over the details of the search and discovery with a Sergeant Jacklin. In the morning we hiked up to the site of the crash with a party of RCMP and about ninety local volunteers, arriving around eleven A.M. You could smell it a mile away. The whole crash area was black. Trees around the perimeter were still licking flame and on the ground bits of seat cushion and clothing were still smouldering.

We did what we could to help the search party with the bodies.

Some were whole and recognizable; of others only charred fragments remained. A number had been entirely incinerated. McQueen himself had been thrown clear and was identifiable. We just tried to separate them into piles, which we tied to the stretchers and struggled down the mountain with. In the end something like sixteen were identified. Some of these were taken away and given separate funerals by their families and the rest were interred in a mass grave in Nanaimo.

You can be very efficient operating in shock. I located a broken snag two hundred feet from the wreck that had been clipped as the plane came in and this allowed us to determine the direction of flight as 150 degrees magnetic. We found the ignition switch panel with both switches on. Both tachometers were recovered and both indicated normal cruising RPM. Two wristwatches were found, one indicating 6:58 and the other 6:59, ten and eleven minutes after our last radio contact with the flight, when the pilot reported his location as twenty miles out of Vancouver. The crash site was over forty miles out of Vancouver.

I interviewed the man who had reported the crash to police, Mr. W. Garrow. He said that at approximately 18:58 hours he heard a large aircraft pass over his house in a southeasterly direction. He couldn't see it due to fog, but thought it was very low. He said one engine sounded as if it were backfiring and as the ship went over three red flares were dropped, descending slowly through the mist for about a minute. The engine noise continued until the aircraft was heard to crash, after which a bright flare lit up the mountainside. Mr. W.W. Walker, whose house was on the road to the crash site, also heard the aircraft go over at about seven o'clock. The noise of the aircraft and the direction of flight being unusual, he ran outside in an attempt to see it. He could follow the course of the aeroplane by ear but couldn't see anything through the fog. He saw no flares dropped but saw a bright flash through the mist several seconds before the sound of the engine stopped, and then a "terrific noise." As he was over a mile distant, the lag between the flash and the sound was explainable. Some other witnesses said the plane circled low over the city a couple of times before flying into the mountain.

We checked our records and determined there were no red flares on FOQ. There were only the white magnesium kind, and they were found still aboard, undischarged. The engines were gone over minutely and as far as could be determined nothing was out of order. In any case the Canso could fly and even climb on one engine.

We had been through enough crashes now to know that people saw all kinds of things that never happened.

We went back and took a statement from our Kemano agent, Harry Lowe, the last person to have seen the pilot alive. He said Doug seemed fine.

"Apart from being cold due to wind, the captain seemed to be in normal good health and spirits when I was talking to him and he remarked that he was glad to be able to get into Kemano on a [windy] day like that when Central BC Airways were grounded. He said it would do a lot of good by showing Morrison-Knudsen that Norsemen were not the aircraft for that job.

"The only other remarks he made were that he would have no difficulty taking off into the wind, which had then reached forty to forty-five, and that the tide was stronger than he expected and the water too deep to anchor in.

"The aircraft refuelled and made a normal takeoff. Subsequent radio messages indicated that everything was proceeding without incident."

They took off at 15:33 estimating Vancouver at 18:10, eleven minutes past grounding time of 17:59. The weather was clear and the pilot probably headed on a direct course for Vancouver, thinking he could make up the eleven minutes by positioning himself to take advantage of the westerly wind usual in this latitude. They passed over Sullivan Bay at 17:35. At this point he found he was forty-nine minutes behind ETA, rather than ahead. He had the choice here of landing to remain overnight at Sullivan Bay, Alert Bay or Port Hardy, of continuing VFR or requesting IFR. Weather conditions were not optimal for landing at any of the available bases so he chose to continue VFR. At 18:24 he contacted Vancouver from abeam Comox with a new ETA of 19:02. He was probably at that time descending between scattered layers of cloud, intending to traverse Georgia Strait at an altitude of about two thousand feet. It was now dark but he had still not requested IFR clearance.

We were flying some IFR and some VFR around this time, depending on the qualifications of the flight crew. McQueen, of course, was not fully qualified, although like most pilots he knew how to use the instruments when he had to. We wanted to go full IFR for night flying but two things were stopping us: DoT was refusing to recognize this particular Canso, a PBY-5A, as an IFR-certifiable aircraft; and the government was stalling installation of a radio beacon near Kitimat.

Comox weather now indicated broken cloud down to a thousand feet and rain, so continuing on to the fully lighted and controlled airport at Vancouver probably appeared preferable to landing in fog and rain at Comox. The last radio report came at 18:48 estimating position twenty miles west of Vancouver and clearing to tower frequency. Ten minutes later he flew into the side of Mount Benson at sixteen hundred feet.

It seemed a great mystery. What was he doing fooling around over behind Nanaimo when he was supposed to be landing at Vancouver? Icing was suggested as a possible factor, but there were no icing conditions that night. The aircraft was well under maximum load and had enough fuel for four more hours flight at cruising speed. But there were some clues. McQueen was running behind time the whole trip. He probably believed he was further south than he was. He was also encountering a wind on his left side, which put him thirty-seven miles to the right of his proper course, something he failed to report and evidently didn't notice in the dark. This positioned the aircraft west of Nanaimo, but since McQueen last reported himself west of Vancouver, that's undoubtedly where he thought he was. Approaching Vancouver from the west you sometimes have to turn right and fly out over the strait to position the aircraft for an approach from the sea, and FOQ struck Mount Benson after having turned fifty degrees to the right.

Nothing was proven one way or another in the subsequent coroner's inquest, DoT investigation, lawsuits or insurance investigations. But we were satisfied that we knew what must have happened. He had been flying by eyeball after dark instead of using his instruments and had mistaken the lights of Nanaimo for Vancouver. It was the same old bushpilot stunt that the hangar crowd figured had been responsible for the Canadian Pacific's Lodestar crash on Mount Cheam in 1942. In that one an oldtime VFR pilot had looked out the window, so the story goes, and decided the little island in the middle of Harrison Lake was the little island in the middle of Howe Sound, then turned into Mount Cheam thinking he was turning over Vancouver. A second section following the same course minutes behind kept their eyes on their instruments and landed safely. But a lot of the oldtimers found it hard to keep their heads down. During the entire darkness portion of Flight 102 McQueen could have used the regular A and N radio range facilities which were adequate to navigate and land by, but he must have looked out the window at some lights glowing through the mist and decided he knew where he was.

It was the worst aviation disaster in BC history to that time. Ours. It was my worst fear realized. It seemed so terribly unfair — we were the people who were trying to enforce safe flying rules using reliable twin-engine aircraft and Baker was the one who was flying overloaded and blind in old single-engine floatplanes, but it was us who had the crash.

One of the people aboard had been Eric Melanson, who had become one of our very few friends in court at Kitimat. He had five children. He had been flying home to help his wife move them to a new home in Vancouver. He'd supported us at the ticket counter as well as in the boardroom and now we'd killed him. The Kitimat Constructors' comptroller was also aboard, along with a business agent for the Labourers' Union, a consulting engineer, a big-name construction boss from the US, the second cook at Kemano, a chokerman, an electrician, a labourer paying a surprise visit to his family and a lot of Kemano construction workers going off on leave. It was just bloody awful. It took me days to come out of shock.

The publicity went world wide. Princess Elizabeth, who was touring the country, extended her condolences. The story displaced news of British-Egyptian clashes over the Suez Canal, where peace had been teetering precariously for weeks, and stayed at the top of page one day after day as more sordid eyewitness accounts were dragged out and more incriminating revelations were dug up. It was like a nightmare that wouldn't stop. It left my head just spinning, day after day. It all came out about the pilot's lack of instrument flying accreditation, his deafness, the aircraft's lack of IFR certificate. It turned out that the co-pilot, Jaginder Johl, had no commercial pilot's licence at all. He was just a baggage handler McQueen had grabbed off the loading ramp to warm the seat. That all came out, plus all the dirt the papers could find about our past flying record. It was just three months before this we'd lost Norseman CF-GRQ on the west coast with the seven people in it, and they didn't waste any time reminding people about that.

The worst thing was having to go down each day after all this public roasting and personal grief and continue trying to operate the business. The DoT came out and seized all records having anything to do with the crash, pending an investigation. I didn't know if this meant we were going to be put out of business or what. I contacted Ottawa and talked to Air Vice-Marshall Cowley trying to find out what we might expect, and the indication was — nothing. The DoT in due course would get around to conducting an enquiry and until then it would be business as usual.

This gave us a bit of breathing space and I decided to try and head them off by taking action of our own. I conducted our own enquiry and cleaned house. John Hatch stepped down from any connection with flight operations and in fact from general manager, where it would come under him. I appointed him marketing manager, where he could get out and look for new business and contact prospective customers, a job he was well suited for. I moved Art Barran up to operations manager and issued a memo reorganizing the operations department by requiring the manager to report directly to me. I put Bill Peters in as acting chief pilot. I then went to good old Grant McConachie at CPAL and asked him for help in smartening our operation up. He was very sympathetic and said how he'd been through exactly the same sorts of things and boy! Jim, he said, no good crying over it—you just have to get busy and fix things up. Could he be of any help? I felt we very badly needed a good man in the flight department. He said he could help temporarily. He had a senior captain by the name of Ralph Leslie who was under six-month suspension for cracking up a Britannia at Hong Kong. We borrowed Leslie for awhile and he was quite a help. He knew all about pilot training and that sort of thing. I also wrote the DoT Technical Division inviting them to come in and scrutinize our procedures.

On the 24th of October, seven days after the crash, I headed for Ottawa and went in to see the Air Transport Board. We spent two-and-a-half hours discussing all the details leading up to the crash and all the expected side effects. Once again I took the opportunity to point out that lack of proper facilities and lack of control over competitive flying had very definitely contributed to the disaster. The board said they'd been in touch with the DoT and, at the present moment at least, no blame was being attached. I gave them an outline of the steps I had taken within the company, making pretty well a clean sweep of our operations department. We were going to control the flying, and to heck with the competition. They seemed pretty satisfied. In fact they were taking a pretty philosophical view of the whole affair, saying, in effect, "Accidents will happen. All airlines have them one time or another and you simply have to survive them." Rimouski Airways had survived the loss of a DC-3 with twenty-nine people aboard in 1948, and they were smaller than us.

Romeo Vachon went further. He slammed his hand down on the

table and said, "You are allowing yourselves to be pushed around too much by this Alcan bunch. Who the hell do they think they are? You should make a firm stand and not back down. The Air Transport Board will stand right behind you!" It was a wonderful little speech. Baldwin somewhat more soberly suggested that it was in our interest to keep Alcan from reacting too negatively to the crash and to this end he would talk to Alcan traffic manager T.C. Lockwood, who was in Ottawa, and assure him the board was satisfied with everything QCA had done. It was typical of Baldwin's kindness and good sense and I was grateful for it. I wrote Dad on October 27:

> Sorry I didn't have a chance to phone you the last three or four days before I left Vancouver but my time was extremely full and did not get home at all in the evenings. Investigations, inquests, and funerals.
>
> I find the official reaction in Ottawa very favourable and feel considerably relieved accordingly. Our remaining worry now is insurance claims and Alcan reaction, both of which it is too early to predict.

We went back to Vancouver and tried to carry on, but it wasn't easy. Baker could hardly contain himself at our misfortune and wasted no time capitalizing. Through the papers he referred to our "terrible accident record" in every connection he could. He got hold of our customers individually and tried to scare them. Baker had a publicity man, a character named Al Williamson, who later went to jail for forging Premier W.A.C. Bennett's signature on behalf of an international fugitive named Harry Stonehill. He was working full time feeding stuff to the *Sun* and they'd come out with the goddamndest headlines two and three times a week.

Baker reserved his best efforts for Alcan, playing up the idea QCA was run buy a bunch of amateurs and alcoholics who should never have been entrusted with the lives of decent people, and whose criminal irresponsibility was directly to blame for the death of their good buddies and co-workers. Feelings were very raw, and he was able to inflame them very easily. The result was we were now practically cut off by Alcan. The whole KKK project area became very hostile. There were threats that we would be refused permission to land in Kitimat, and cooperation from shore dropped to a new low.

All this time I was fighting a rearguard action from our own unions, trying to get them to hold down demands while we fought Baker for the big Alcan prize, explaining to them the additional disadvantage of being the only unionized airline in the country that didn't get mail pay. As our position at Kitimat worsened and nobody else seemed willing to help, it occurred to me that the unions might be able to do something. I pointed out that ninety percent of the KKK traffic was comprised of union members—why couldn't our union people apply through the labour organizations to have their brothers boycott Baker, or at least give preference to the unionized operator? Our union heads saw my point right away and said they would get right to work on it. But nothing came of it.

On the other hand A.R. Eddie, executive vice-president of the Canadian Airline Pilots Association (CALPA), provided us with a very strong letter of support following the crash. It said, in part:

> For some time we have watched with concern the flying operations being conducted on the west coast, and would respectfully offer the results of our research into the matter, with particular emphasis on the Vancouver–Kemano–Kitimat operation.
>
> We understand that largely through the insistence of the firms interested in the aluminum project, the Kitimat service was thrown open to competition, no protection being accorded any single operator...
>
> Our experience in aviation goes back to the early days when there was little or no regulation of air routes, with the result that several operators frequently tried to serve the same general area. This nearly always resulted in a dog-eat-dog situation wherein...pilots who would not take chances were looked down on by those who were in a position to place flying contracts...rarely did a season pass without an aircraft going through the ice because someone had attempted to hurry winter operations, or an accident occur through people attempting to fly in bad weather...the wonder is that more people were not killed.
>
> When [in 1944] the Air Transport Board adopted a policy of restricted and sensible competition everyone with regard for the welfare of aviation

breathed a sigh of relief. However, from here it appeared to us that the situation on the west coast was likely to degenerate into much the same rat race that obtained in the thirties, with all concerned faced with the alternative of placing immediate economic advantage ahead of safe practice, or else withdrawing from the area. The prospect of thus returning to the dark ages of aviation...in an area which presents extremely difficult flying conditions was not and is not at all welcome.

With particular regard to the Kitimat run, we believe that another danger exists in that there are no facilities at this place for safely mooring an aircraft for the night...

We believe that the following action would go far toward rectifying the situation:

> 1. The granting of the route licence to an operator who will provide equipment and trained personnel to safely operate the service, and who is protected to a great enough extent to justify his expenditure on equipment and training...
>
> 2. Construction of a ramp at Kitimat to enable aircraft to be drawn up on shore when conditions warrant laying over there.

The letter was sent to the relevant government departments and to Alcan, but had no more effect than our own protests did. Finally I decided to cancel all KKK flights for the duration of the winter of 1951–52 and see how things looked come spring.

Meanwhile, we had this investigation. The Department of Transport sent an investigator out to decide our fate. Why they sent the particular man they did, I have no idea. I felt that both Cowley of the DoT and Baldwin of the ATB had been sympathetic when I'd visited in October, but they couldn't have picked a worse guy from our point of view. I will call him Ballsup.

I received a call one night about two o'clock in the morning and it was Baker, very drunk, calling from his party suite at the Ritz Hotel. It woke the whole house up and Glenys was mad.

"Well," he said, "we've finished the meeting."

"What meeting?" I said.

"The meeting on QCA," he said. He mentioned this investigator's name and said he had him there with him. To this day I don't know if he really did, but I could hear voices in the background, all as drunken as his. They had decided between them, Baker said, that henceforth no Queen Charlotte Airline aircraft would be permitted to fly more than one hundred miles from its base. I didn't believe him for a minute. And even if he was telling the truth, and they had dreamed this up in their drunkenness, the idea wouldn't last through the sobriety of the following morning. So I thought. But I was wrong. In the morning came notification from the DoT that this was their decision. No more than a hundred miles from base.

That stopped us. The implication was, I suppose, that we weren't to be trusted maintaining proper procedures over long flights like Kitimat–Vancouver, because of the trouble McQueen had got into. But it was physically not possible. Our Vancouver–Tofino flight was 121 miles. Vancouver–Zeballos was 140 miles. Vancouver–Alert Bay was 192 miles. Vancouver–Kitimat was 392 miles. Vancouver–Ocean Falls was 344 miles. Prince Rupert–Stewart was 120 miles. No one else could fly those routes. If we were to be grounded, the thing to do was lift our licences and hold new, open hearings for applications — in which case Baker would not be assured of getting all or any of the licences. Otherwise we had to be allowed to continue. But Baker's intention was to cripple QCA so we would be forced to sell our licences to him for nothing and Ballsup's decision couldn't have suited Baker's purpose better if he'd written it himself. I had to fly back to Ottawa and after three days of alternately pleading and threatening I had the restrictions removed from our main schedule services to Tofino, Nanaimo, Comox and Sullivan Bay, although Ballsup succeeded in keeping the hundred-mile limit on our Stranraers and Norsemen, as well as his ban on all IFR night flying (while at the same time requiring that all QCA Canso pilots have IFR accreditation), and a particularly dirty provision which placed Kemano off limits to our Cansos — effectively reserving that prime traffic point for Baker's Norsemen.

THIRTEEN

Aftermath

WE'D HAD ACCIDENTS BEFORE, even serious ones, but this time around it devastated the company. Traffic plummeted on all routes. In September 1951, the month previous to Mount Benson, we had a profit of $20,543. The month of the crash, October, the profit turned to a loss of $32,062. The next month we lost $68,608; the next $58,977. We didn't have the capital reserves to withstand this kind of beating for very long. And when would it end? What would it take to turn it around?

The crash had weakened QCA to the point we could no longer kid ourselves about the threat that Baker posed to us. It was now all too possible that with his financial backing from Springer, his solid foothold at Kitimat and his hold on the newspapers, he could indeed end up winning the battle for the coast. It became a day-to-day vigil to see whether QCA could pull out of its dive, with the whole aviation community anxiously looking on.

The accident wasn't the only reason we were flying empty planes. Baker had escalated his poaching. Whether he had planned this from the beginning or whether he got the idea as he went along I'm not certain. I'm inclined to think the latter. In any case, here he was day after day flying over our main route from Kitimat to Ocean Falls to Alert Bay to Vancouver, hauling freight up and coming back empty. And all the way down the coast, here's planeloads of juicy-looking QCA passengers waiting for our sched flight to come in. For Baker this was too good to pass up. We soon found, if we were due to take off at three o'clock from Ocean Falls, Baker's man would come along at two o'clock and say, "Anyone for Vancouver?" And whatever the fare was, he'd cut it in half. And once again, he could charge any price because his trip had already been paid for. Anything he picked up on the return was found money. As Baker warmed to the possibilities of this wide-open kind of piracy, he started to monitor our radio frequencies to find where we had people waiting, and he'd drop a plane off at Alert Bay, Sullivan Bay, at Minstrel Island and eventually even Tahsis and Zeballos over on the west coast. We'd had this for a year already on our KKK end, but now we had it over our whole route system. It kept up month after month. We kept documenting the CBCA violations and sending them to the Air Transport Board, and eventually they began to pay a little attention. I think they sent Baker a few enquiries and warnings.

I didn't know it then, but someone on the inside was slipping him copies of our letters of complaint. They were found in his personal papers many years later. This would have given him a tremendous advantage in being able to head off investigation and probably explains some of the Air Transport Board's slowness to act on our reports. Evidently during this whole period Baker had one or more agents working for him, but I don't know who they were. Duncan McLaren wrote up most of our protests, and he has apparently voiced the opinion the leak was at the Air Transport Board end. This could be, but there was one other leaked document that pointed to our own end. This was a tongue-in-cheek report I'd written for the benefit of our board after visiting Kemano in 1952. Canada had never seen anything like this Morrison-Knudsen bunch. They were marching around up in north-central BC like a division of the US army. The bosses wore special hardhats with stars on the brim indicating their status. I still found it very hard to sort out just who was in charge and in writing up my report I made rather a joke of this, quoting various individuals and noting whether this was a one-star, a two-star or a three-star opinion. It was never meant to go beyond our directors, but I learned soon after that it somehow got back to Kitimat and all the brass up there were handing it around, probably steeling their determination to prevent our planes from ever taking a full load again. It was a public relations master stroke by somebody. It completely undid all the work I'd been doing trying to weave our way back into their good graces since Mount Benson.

Now more than ever I saw that we had to have those crucial airmail revenues to survive. In March, 1952 I appointed a management committee consisting of Al Houghton, Dunc McLaren and John Hatch to look after things while I was away, and sallied once more to Ottawa, this time armed with statistics to back up my demands:

Company	1952 Mail Pay	Mail Pay as % of Passenger Revenues
TCA	$5,741,000	20.02
CPAL	$1,380,651	27.12
MCA	$ 159,093	61.95
CNA	$ 51,837	34.68
QCA	$ 14,785	1.47

The average ratio of mail pay to passenger revenue across the country was 20.72%, which should have entitled us to $192,767 annually. How could they defend such outrageous discrimination? I went into the post office to meet the new postmaster general, a man named Cotes, but was directed instead to a man named Tedford, who was director of air services. This was the first time anyone had thought it necessary to introduce me to this obviously crucial figure, but he knew all about us. He was very pleased with the services we were providing on various parts of the coast and even acknowledged the merit of our latest proposal for an all-points service on a frequent basis with two hundred pounds of mail guaranteed at air freight rates. This he admitted was a far better deal than they were getting from most operators. I made our case as strongly as I could and left him sifting through a pile of backup documents two inches thick.

Three months later I was back in Ottawa again, still with no gains on any front. We had tried to resume our KKK service, but Alcan had out-and-out refused to let us land at Kitimat. Losses continued to pile up. I made all the usual rounds. I went to see MPs Jim Sinclair and George Applewhaite and they were, in my journal notes at the time, "incensed and appalled" at Alcan's behaviour towards us. The bureaucrats at the Air Transport Board were as usual full of big talk about what we should do in the future but silent on simple answers to our current problems. I begged them once again to stop Baker's unbridled piracy and make Alcan install a proper beaching ramp at Kitimat, but all they offered to do was write a letter. They didn't want to dwell on such mundane details; they had been talking over our situation and had come up with a master plan for us.

The way out of our troubles, they advised me, was to look at ourselves and realize where our strength lay, then make the most of it. And our strength was in our development as a high-grade operator. We were well on the way to becoming a proper Class A carrier when Baker came along and side-tracked us into a contest over low-grade business. Now they wanted us to get back on track and complete our development. We should sell off our small aircraft, upgrade our scheduled runs to Instrument Flight Rules using twenty-one-passenger Douglas DC-3s and become an exclusively scheduled carrier, leaving the Class B and C business to the others. We would have all the larger aircraft and fly mainline, or "spine" routes. Baker would be restricted to small single-engine stuff, fighting it out with the other small operators for the "feeder" business. If we did this, I was given to understand we could expect to receive fully protected Class 1 routes such as Vancouver–Port Hardy and Vancouver–Terrace.

Baldwin warned this would be possible only with equity financing. He wanted us to re-finance the company by getting some backers of real substance and selling controlling interest to them. He referred me to an Ottawa lawyer named Jimmy Wells who had successfully arranged a re-financing plan along the same lines for Carl Burke of Maritime Central Airlines. I did see Wells and he did say that he would help set us up with third-party financing when we were ready, but when I brought the idea up at QCA directors' meetings I couldn't get it past Lando. He had been taking shares in lieu of fees over the years and had become QCA's largest shareholder. Now he didn't want to sell off control. He always maintained the way to business success was to make the thing finance itself as it went along.

I agreed with the Air Transport Board that our salvation lay in the direction of high-grade service. I had always seen scheduled flying with larger aircraft as the future anyway—I just had trouble bringing it about. Now that the ATB was apparently ready to give us the Class 1 routes it seemed imperative we make our move, but right then it would take the one thing we didn't have—money. Our bank debt was edging up over three hundred thousand and Standard Oil was demanding action on the seventy thousand we owed them. We owed a further fifty thousand to small creditors, most of it dangerously overdue.

Airmail pay had been my main hope of easing our financial squeeze, but now I turned my attention to the government-assisted loan program we'd been working on through the AITA. Banks wouldn't loan money on aeroplanes. They couldn't, under the terms of the Banking Act. The problem had to do with the nature of the business. If you bought an aeroplane you might register it in British Columbia. But six hours later it could be in Manitoba. It could be anywhere. It was just damn poor security. It slipped through your fingers. Without banks fully behind them, operators had to finance their planes through finance companies and high-risk lenders like Captain Hubbard, a shady character who lurked in the aisles at AITA meetings offering loans at rates up to forty-five percent.

For several years now I'd been making noise at AITA meetings about finding alternate means of raising money to buy aircraft. Our first idea — for a government-funded transport pool which could be used to fly troops in wartime — never got off the ground, but we kept casting around and eventually came up with something much simpler. It was Bud Lando who pointed it out to me one day. The Industrial Development Bank (IDB) had been formed some time previously by the federal government and it was doing great things providing equity capital for traditional industries, but they were up against the same limitations as chartered banks when it came to financing aircraft. At the next AITA meeting, which happened to be in Sun Valley, Idaho, I stood up and said, "Why don't we just ask the government to change the regulations governing the IDB so they can finance aircraft through it?"

Lionel Chevrier was transport minister at this time and he was in attendance. Howard Cotterell was AITA president and after the meeting ended, he, Chevrier and Bob Redmayne called me in and Chevrier said, "Look boys, if you people will roll your sleeves up and go to work on this, form a committee and make a formal recommendation through the Air Transport Board, why, I would be all too glad to put it up to cabinet." Cotterell asked me if I would chair an aircraft finance committee to gather data and make a formal presentation, so I was nabbed. The other members were Milt Ashton of Central Northern Airline in Winnipeg; Doug Kendall, a young fellow from Photographic Surveys in Ottawa; and Karl Springer, Baker's partner in CBCA. We held our first meeting right there. Another one we held in Winnipeg. I don't think we had more than about three.

The sparkplug was Doug Kendall. He was a tremendous help. Milt Ashton said he'd tried it all before and it wouldn't work, but he at least paid attention. Karl Springer was something else. When we were having our final meeting, getting the proposal ready to submit to the AITA annual meeting in 1951, Springer slumped down in his chair so drunk he couldn't move. At the end of the meeting we went away and left him sprawled with his chin on the table. He took no apparent interest in the thing — which was amazing, because at that time Baker's need for financing new aircraft was almost as desperate as ours. Anyway, we finished our proposal, submitted it to the meeting, they accepted it, they applauded it, they congratulated us and Chevrier said "Boys, you've done a wonderful job. I'll take it from here." It was approved by Order in Council, for the first time creating a major source of capital funding for Canadian airlines.

Howard Cotterell of Trans Canada Airlines wrote me a personal letter saying, "To me the most encouraging thing that has happened since my first connection with the association was the way you people took the ball [on the IDB] and came through. I will stand up to anybody and say, regardless of the shortcomings of the panel, in retrospect it is the best thing that has happened in the association in the last five years, and probably since the inception of the organization."

Nothing but accolades in every direction. I was feeling pretty pleased. Back home at QCA we lost no time in preparing our own application, confident that after all this work we could get the two hundred thousand we needed to finance our big move into DC-3s.

There was a long delay.

Then the IDB turned our application down.

The reason was no different than the one banks always give. They wouldn't loan us money because we didn't have enough money. If we'd had lots of money they'd have loaned us some more. They felt we didn't have enough equity capital. John Baldwin had harped on this of course, and said well, I told you so.

What we didn't know was at the same time Baker had applied and been given $234,000. Once again he'd done an end run on us. The man most responsible for assisting on the loan was the IDB's west coast manager, who was stationed in Vancouver. As this man later told author John Condit, "On the basis of what are normally considered good banking standards, the [Baker] loan didn't look so good." But there were other considerations to which normal banking standards apparently didn't apply. One, according to the IDB man was, "All this business at Kitimat was of very great importance to the province of BC and this sold the bank on the need." Well, QCA was in the Kitimat business, too — in fact we were the government's designated scheduled carrier there, but the IDB didn't see that as a reason to bend the rules for our benefit.

What may have helped Baker more was his personal relationship with the local IDB manager himself. While I spent my time locked away in smoky committee rooms digesting financial information for AITA, Baker had been at work on him. They went to the races together, they talked and the guy became what I would call a real Baker booster. One of the things that apparently won the IDB over was a cameo appearance by lumber tycoon H.R. MacMillan. Baker had been cultivating MacMillan ever since he'd flown him into Cold Fish Lake on a hunting trip in 1949. We'd had our chances with H.R. MacMillan going back five years before that, but all we'd managed to do was get into a squabble over his bill. Baker came out of his MacMillan charter with a lifelong supporter. According to his wife Madge, Baker and MacMillan would afterwards "sit talking for hours."

MacMillan didn't co-sign anything to back Baker's IDB loan. Being the wealthiest man in western Canada, he didn't always have to sign things. All he did was walk into the IDB office and say, "The only reason I am here is not because I can discuss the merits of the financial proposal, but because I can tell you that you can depend on Russ Baker not only for honesty but also for his ability to do the job, and that he will carry out what he says he will do." Apparently this sweet message from on high "helped to carry it [Baker's loan application] through the whole bank."

The money was supposed to be used to buy two new "King Beavers" as the de Havilland Otter was at first known, but the most notable sign of new wealth on Baker's part was a fourteen-room mansion on the highest hill in West Vancouver's posh British Properties. We always knew he had lots of money he didn't make himself, but we thought it was all coming from Daddy Springer. In actual fact, Springer about this time pulled $165,000 of his own money out of Central BC Airways, apparently replacing it with the IDB money our hard-working committee had dug up while he slept.

The collapse of our IDB hopes left us once again placing our faith in mail pay, but all through the fall and winter of 1952, as our revenues dried up and our bank debt ballooned, the post office bureaucrats stalled and waffled and dithered. To add to our financial miseries, the summer of 1952 brought prolonged layoffs and strikes in both the logging and fishing industries, leaving our prime revenues at rock bottom when we needed them most. Up to the end of August, we had lost $165,000. We were now heading into what was traditionally our off season, and we could only expect our losses to worsen.

We knew we couldn't weather it. Bankruptcy was now staring us full in the face. Unless something happened to alter the picture drastically, QCA would be gone by spring and Baker would have the coast to himself, just like he wanted. I don't know which bothered me worse.

On October 2, 1952 I presented the directors with a restraint program. We had already laid off all staff we could possibly spare, but now I proposed we make radical service cuts as well. At the top of the hit-list was Vancouver–Nanaimo, which had lost $28,000 between January and August. Others were Ocean Falls, Seymour Inlet and Alliford Bay, which had lost some $45,000. Comox had lost $37,000, but it was our showpiece Class 1 service and I didn't want to give that up. I suggested we concentrate on maintaining the Comox–Powell River triangle, Tahsis and the west coast of Vancouver Island, Alert Bay–Minstrel Island, Kitimat and the Prince Rupert base serving Alcan, Stewart and Masset. Besides allowing us to lighten the payroll further, the reduced service would allow us to sell off two Norsemen. If we sold the five remaining Ansons as well, we might have enough out of the lot to buy the two DC-3s we needed to embark on the brave new future I had discussed with Baldwin. At the same time we would try to raise $100,000 by sale of stock in the company and form a brain trust of all our top people to study the possibilities of making another proposal to the IDB which they might find acceptable. Any funds raised in this way would be used to satisfy the more impatient creditors. The rest would be pegged and put on a COD basis until we were able to pay them off.

The bank was getting very panicky and I had to make a special trip to Montreal to walk them through our survival plan so they wouldn't close us down. I left them a *little* happier than when I found them. They were very eager to have the IDB share some of the QCA debt load, but when I went to interview a Mr. Noble at the IDB about our chances on a revised application he turned me down cold. He said they wouldn't split our financing with another bank and we should stick with who we had. When I told Baldwin about this, he was shocked. The whole idea as he'd understood it had been to have the IDB cover areas the chartered banks wouldn't.

In November, 1952 I was back in Ottawa again to visit Tedford, the air services man at the post office, but all he had done in the intervening six months was think up a new excuse for doing nothing: he said he couldn't provide any new mail pay to QCA as long as the

Union Steamship Company was technically still in service, and at this point it was still weakly hanging on with greatly reduced service. It was about the most dubious position I'd heard from the post office yet, but I couldn't get him to budge.

We were still fighting the restrictions imposed by Ballsup so I went in to see Air Vice-Marshall Cowley at the DoT to determine what progress had been made toward having them lifted. He wanted to know whether I thought they should just lift them, or keep them on but apply them equally to all carriers. My trip journal records that my answer to this riddle was: "Yes." I didn't care which way they went as long as it put us on an equal footing with our competitors.

By December, 1952 still nothing had improved on any front — mail pay, DoT restrictions or Kitimat business. Finally I went to see Jimmy Sinclair and said look, this is an emergency. I showed him our current financial statements projecting a loss on the year of almost half a million. I said, "Our payroll is coming up in two weeks and we aren't going to be able to meet it. If you people don't do something, the west coast is going to have a major collapse of its air services and your government will be directly responsible."

Sinclair at this time enjoyed a very high reputation in Ottawa, although the senior minister from BC was solicitor general Ralph Campney. Campney was a prominent Vancouver lawyer and his firm, Campney, Owen and Murphy, handled Baker's Air Transport Board applications and CBAC's interventions against ours. Campney was resisting any move to assist QCA. Sinclair got quite hot about this and told me the way they'd divided BC up between them was Campney took the cities and Sinclair took the country. The airlines were in Sinclair's territory and he felt Campney had no business interfering, but if he wanted a bust-up, well, this was a good issue to have it over.

Meanwhile we had started taking the drastic kind of measures you have to take when you're faced with bankruptcy. We sublet all of the Class 3 and 4 charter routes to BC Airlines and leased some Norsemen out to our good customers at low rates, sold some, mothballed others. Then came the hard part for me: facing the fact that we had to let many of our staff go. We had already cut about fifty or sixty from our maximum level of 250, which wasn't so hard because it didn't affect our core people. Now we had to go down the list of our best and most loyal employees, the people who had helped make the company and who *were* the company in an important sense, and cut a good portion of them loose. To make the task just as unpleasant as possible, it was only about two weeks before Christmas. Agonizing over who to keep and who to chop just about drove me around the bend.

Finally, in Ottawa, Jimmy Sinclair began to see some light. He called me up on the 17th of December, 1952 and told me to get out there before ten o'clock the next morning. I was still struggling with the layoff problem, but Bud Lando stepped into the breach and said, "Look, you've got enough to worry about. You leave that to us. You get off to Ottawa and get the money. We'll look after the cutback." And boy, they did. He and McLaren just went down the list blindly slashing off names until they had a hundred. It was murder. Even loyal, long-serving Norm Landahl got the axe. I was appalled at some of the mistakes they made, but I couldn't complain since I'd been unable to finish it myself.

In Ottawa I was met at the airport by Angus Morrison, secretary of the AITA, and whisked right down town to the Number 1 Temporary Building in which C.D. Howe's office was housed. The meeting of the special committee on QCA — Campney; C.D. Howe; Cotes, the postmaster general; Lionel Chevrier, the minister of transport — was just over. They told me Jim Sinclair had brought the whole matter before the cabinet the day before and had done a fine job showing what a spot west coast transportation had come to with the steamship company in collapse and the leading airline near to it. When he got through, Prime Minister St. Laurent said he was astounded and shocked at the discriminatory mail rates given QCA. "QCA must not be allowed to fall," St. Laurent said, and directed the special committee to find a solution.

Lionel Chevrier thought help should be given in the form of normal mail pay, which would be a permanent solution. C.D. Howe was against this; he favoured a one-time subsidy. Mr. Cotes was against mail pay but in favour of some kind of assistance. Someone had suggested a loan. Campney was neutral. To nobody's surprise, C.D. Howe prevailed. They granted QCA $25,000 a month for the six winter months. QCA had to undertake to maintain all Class 1 and Class 2 protected services, and agree not to start up any new routes or services or any work that might lose money. The situation would be reviewed in August, 1953 and if QCA were not in a sufficiently improved position by that time the board would have to consider either transferring our licences to some other operators or a forced sale of the company.

We had our subsidy, but Baldwin cautioned me there would be

considerable delay before we actually got our hands on any cash. I told him this was a problem, as I was quite sure I couldn't talk the bank into extending our credit any further, no matter what assurances I gave. Baldwin then did something that opened my eyes considerably on what the phrase "behind the scenes" really means. He picked up the phone, called C.D. Howe, Howe called the bank president, and bingo — credit extended. No papers to sign, no regulations quoted, no financial reports asked for, no fuss, no muss. It only took a few minutes. This was how they did things at the top. It was a bit chilling to see up close, but it was nice to be on the right side of it for once. As minister of trade and commerce, Howe was then at the height of his power, the virtual dictator of the Canadian economy.

Before I left, Baldwin and I went over our financial projections one more time with the subsidy added in. It would only cover half of our losses on the year, but it would allow us to meet the payroll and get us through to the high revenue months of the summer, and I still thought we could make $250,000 before the winter of 1953. He said he would be happy with $150,000, but he made it very clear the board insisted we now get DC-3 equipment without delay, get rid of our small aircraft charter work and become a Class 1 scheduled operator exclusively, like TCA and CPAL. I told him the transfer of our Class 3 and 4 routes to BC Airlines appeared to be working out well, which pleased him mightily. The board considered this move of our non-scheduled licences to BCAL one of the brightest things that happened on the coast.

I got back to work just a day or two before Christmas, feeling like I'd been through a wringer but feeling fairly pleased, too. I'd gone to Ottawa with the sole purpose of getting that subsidy, and by golly, I'd got it. Somehow the office seemed strange. People sort of looked at me and there wasn't the usual chorus of "welcome home, boss!"

I couldn't quite put my finger on it, but then we'd had that devastating layoff and it was bound to have some effect. I went on into my own office, hung up my coat, took off my overshoes and sat down. There was a lot of mail to be looked at that the secretary had got ready, but two letters stood out. They were placed right in front of my chair. QCA envelopes without stamps. I opened them both. The first was from my executive assistant, Duncan McLaren. It read:

Dear Sir:

It is with considerable regret that I ask you to accept this letter as my formal resignation from employment with Queen Charlotte Airlines Ltd.

During the past seven months, I have found it an extremely difficult task to handle my duties due to the fact that I have lost confidence in the management of the company. In addition the continuing critical financial situation coupled with various management problems is such that I feel I can no longer accept the responsibility for the safety of flight operations.

I think you will agree that employment under such circumstances is impossible and, in the light of the foregoing, I believe it would be in the best interests of both the company and myself if this resignation is made effective immediately.

The second letter was from our controller-treasurer, Alan Houghton, also resigning effective immediately. You can imagine what a shock this was. I had come to rely very strongly on both of these men and took them into every confidence. They knew every detail of our past business and future plans, including the latest developments in Ottawa. I had kept the office up to date on the subsidy negotiations and telephoned the news as soon as I knew it was going through.

There was no sign of either of them in their offices. I think I probably tried to enquire where they were but got nowhere until a young fellow from accounting asked if he could come in to see me. A nice young chap. He looked very scared and asked me if I knew what was happening.

"No, I don't," I said. "Can you tell me?"

He told me that Houghton and McLaren had been going around to people asking them to tender their resignations. McLaren had then gone around to the pilots, he said, and at this very moment he was over in the hangar talking to the maintenance men. The story the two of them were giving, according to this young fellow, was that if they would all resign, Russ Baker would step right in and take over with no interruption in pay. The only difference would be that instead of QCA over the door it would say CBCA.

I got on the phone and phoned our maintenance boss. He said,

"Yeah, they're over here, but what can I do? They've got the authority."

"Not any more, they don't," I said. I told him they had resigned and I wanted them off company property immediately.

The very next thing I did was call an emergency meeting with every employee who could be found on the premises. I reported on the success of my trip to Ottawa and said that the payroll would be met and the future of the airline was assured. I thanked those employees who had remained loyal and named the ones who had resigned.

There were a couple of disturbing things that had taken place in my absence. We normally gave all employees a Christmas hamper, but this year we hadn't been able to. Baker had apparently found out and circulated a memo to all employees — how he managed it I don't know — but he told all employees that if they came over to CBCA on Christmas Eve he'd provide them with a Christmas hamper and turkey, as he understood QCA couldn't afford it. The other thing that had transpired was that Baker had circulated a memo to all our pilots and flight crews directing them to contact CBCA offices to get measured up for CBCA uniforms. I'm not sure that CBCA even had their own uniforms at that time, but it didn't cost Baker anything to measure someone. Several people did go over, including Art Barran. Art was one of our most senior employees. It was very upsetting and very confusing to the rest of our people.

Anyway, I had to straighten all this out and I also had to check in with Bud Lando and Frank Griffiths. Neither they nor my assistant Hal Suddes knew a thing about any attempted palace coup. Whatever had happened had apparently happened behind their backs. Frank Griffiths felt somewhat embarrassed about Houghton because it had been he who'd originally found him and advised us to bring him in. I had to bear the blame for McLaren. We decided that we should be more careful when it came to choosing replacements, but in the meantime it was arranged that Frank Griffiths and Bud Lando would come out every Friday morning at ten o'clock and join with me as a management committee to deal with all immediate matters, mainly financial but also operational, and in that connection we would call in Hal Suddes. We pulled ourselves together very quickly after this and actually had a much smoother running operation from then on.

Suddes in particular was a godsend at this point. He had joined us in June as assistant general manager after serving as AITA secretary under Bob Redmayne. Before that he'd been involved in the original design and manufacture of the Norseman aeroplane with Bob Noorduyn. He was very experienced, very loyal and a very good administrator, and he helped no end in putting things back in shape. To replace Houghton, Frank Griffiths found a very level-headed, experienced accountant by the name of Reg Richards, who in turn brought in an assistant named Fred Nicks. Between the two of them they reorganized the entire accounting department, reduced the staff by five and got things running on a current basis. Fred Nicks was still working with me as my company controller when I retired thirty years later.

After an interval of several months McLaren showed up working for Baker. Houghton we lost track of but McLaren stayed on as Baker's right hand man for a number of years. I wondered in retrospect how long McLaren had been contemplating his jump from QCA to CBCA and how it had affected his handling of the highly strategic work he was called upon to deal with for us.

McLaren has since stated that there was no attempt to coerce our employees and he did not know Russ Baker, Karl Springer, or anyone in the CBCA organization, and had no contact, direct or indirect, with them or their company until the spring of 1953, after he left QCA.

FOURTEEN

Shovelling Smoke

WITH THE SUBSIDY, we'd fended off our financial crisis temporarily, but I was fast running out of allies. Lando and Frank Griffiths worked hard on the management committee but they left the flying decisions to me. Hepburn was still around, tinkering with radios and reading letters upside-down, and he knew so much and had so much interest in the company I found it hard not to call upon him from time to time. The bad feelings surrounding his fall from grace seemed rather insignificant compared to what we'd been through in the meantime, and finally I brought him back as assistant operations manager. Johnny Hatch made an excellent marketing manager, but Mount Benson had been a terrific personal setback for him and I guess it was too much to expect he would stay on indefinitely. In August, 1952 he quit to go into business on his own—the trucking business. It was an emotional shock to see him go finally, and I meant it when I wrote in my letter officially accepting his resignation, "Should the matter ever appear desirable to you, I can assure you again that I will always be very happy to discuss the possibility of your rejoining the Company at any time."

But Johnny's aviation days were over. He made a moderate success of his trucking business, married and remarried several times over the years, and in 1988 was ranching in California, raising a buffalo-shorthorn cross known as *beefalo*.

Back at the airport, we were going ahead on the Air Transport Board–ordered plan to convert exclusively to Class 1 IFR-type operations. Our hope of raising equity capital through share sales and the IDB was a flop, but we did very well selling off surplus equipment. Douglas DC-3s were being unloaded by the big airlines for sixty-five thousand dollars and we had no trouble cashing in enough spare Norsemen to get our first one, CF-HCF, on a lease-purchase deal early in 1953. We were rather nervous about making the leap into real airliners, but the DC-3 was one of the most user-friendly aircraft ever made. It was an operator's dream compared to flying boats, in terms of reliability, maintenance and payload, although of course it was limited to routes where there were thirty-two-hundred-foot airstrips.

We decided it would now be safe to let go the last five of the old Ansons we'd been using on our scheduled wheelplane runs to Comox, Tofino and Powell River and advertised them for sale. They were now *really* obsolete and for a long time we received no interest, then we got a bite. It was from Ecuador's national airline, Aerovias Ecuatorianus CA. A little fat guy in a uniform covered with brass buttons and medals and a great peaked cap came up from Quito to make the deal. Captain Luis Arias Guerra. I believe he was actually the president of the airline. He was a real character. Just like out of a comic book. We were negotiating over some detail and he decided he had to make a phone call. So he got on my phone and got through to Quito and got the president of Ecuador on the line. He sits down in my chair and twists his big black moustache and starts in with this machine-gun Spanish: yadda yadda yadda, yadda yadda yadda, yadda yadda yadda. It goes on and on. Yadda, yadda, yadda. I haven't a clue what's being said, but it sounds like they're reviewing the entire history of the deal, maybe the entire history of Ecuador. Yadda yadda yadda, this guy keeps going on until finally, Si! Si!, and bang! He hangs up. Then he turns to us:

"He say no."

It took him twenty minutes in Spanish to say no, on my long-distance tab. But we eventually got everything ironed out, painted the old plywood fuselages up with Ecuatorianus insignia and sent them down to Quito with crew. They worked fine down there, but they kept putting more kids and goats into them and flying higher up into the Andes until they wouldn't stay up. They stacked them all up in a matter of months. But we turned a damn good buck on the deal. We got forty thousand for the bunch of them, which was more than we needed to get into our second DC-3, CF-EPI.

Ever since Mount Benson we had been without a permanent flight operations manager. We had hired a new chief pilot I hoped to build around, Captain Bill May. Bill had sixteen years with BOAC, eight of them as chief instructor and officer in command of training. He'd also been a bush pilot in northern Manitoba and had over thirteen thousand hours piloting every type of aircraft, large and small. I thought I would be able to promote him to the flight operations job, but Bill wasn't taking hold like I'd hoped. We had given Art Barran another try running ops, but Art never worked out for long. The other pilots resented him, and as I've mentioned before, even though he'd been with us since 1946, he was not loyal. He was very critical of management, so it made it hard to give him any degree of authority in spite of his skills. I got Duncan McLaren to take over from him, but when McLaren jumped ship I looked around our outfit to try and find somebody who would take over permanently, a good solid type, and my eyes came to rest on Sheldon Luck. Sheldon was an oldtime Canadian Airways pilot, he'd flown the west all his life, he was an excellent pilot in every way and I called him in. We had a long talk, I told him the problems and asked him if he would take over in ops. He was the man we needed.

Sheldon was very sympathetic. He said, "Jim, I know. You need somebody. You need somebody bad. But I don't want any part of it. All I want to do is fly an aeroplane. I don't want any responsibility, I don't want any authority, that's nothing but a headache. I just want to fly."

And the bugger wasn't lying. He did want to fly, and he went on flying. He flew everywhere and everything. Years later he was chosen to fly one of the royal parties around BC. He didn't retire until the 1970s, when he was an old man and a true flying legend.

But before Sheldon left my office he thought for a minute, then said, "I do know somebody, though. I think he may be just what you're looking for."

He gave me the name of Eric Bendall. I'd never heard of him, but Sheldon was quite sure that if we could get him our troubles would be over. At this moment Eric was renting a place in White Rock, trying to decide if he liked the west coast well enough to settle down and look for a job. I got his phone number, called him up and he came over for a talk.

I didn't know quite what to make of him. He didn't have the suaveness of many top pilots. Eric was as direct as a poke in the eye. He seemed to have an awfully black-and-white approach. Flying questions that appeared complex to me he seemed to think were simple. When I tried to explain the delicate negotiations that went on trying to get the pilots' union to adopt new procedures, he would say something like, "If they're not prepared to do what they're told, they shouldn't have the job." I really didn't think this kind of talk would get very far back at the hangar.

I laid all our problems in front of him just as black as I could paint them. They made an impressive array, one you'd think would make any prospective flying executive at least want to go home and think things over for a week, especially someone like Eric who'd never been anything but a common line pilot before. But that wasn't Eric Bendall.

"Well, I can see what you need," he said in his blunt way. "I'll take it but I'll tell you this right now. I want complete authority to do anything I figure needs doing. I don't want anybody getting in the road and that includes you."

I asked him if he didn't want to mull it over a little but he said no, he'd come to work the next day, January 7th, 1953. I offered him six hundred a month and a Standard Oil credit card, assuming he would ask for more. It was much less than he'd been getting on his last job. He shrugged and accepted.

Next morning, in walks our new operations manager. I didn't have a chance to warn anyone; I had to issue a memorandum announcing his appointment after he was already at work. The first thing he does is call the pilots in for a meeting and tell them everything is going to be done different. Enough of this bush league stuff. They're all going to have to go back to school and renew their qualifications. Everyone is going to have to go through for full instrument flight rating, and those who can't make the grade will be assigned to light planes or let go. This retraining was going to require lots of after-hours and weekend time with little or no extra pay. Everything was going to be done airline-style, by the book, and anyone who bucked was down the road.

Immediately, he had the union up in arms. No way were they going to be treated like a bunch of naughty schoolkids by some outsider who didn't know one end of the BC coast from the other. They wanted their working conditions and their job classifications left the way they were. Let management get its own house in order, then come talk to them. But Eric was ready for that. "You all know damn well what trouble this company is in," he said. "One more prang and you can *all* kiss your jobs goodbye. I'm here to restore public confidence, and the only way I can do it is by curing all your bad flying habits and upgrading the lot of you to IFR, starting immediately. Once we're in good shape, then we can talk union. But right now we're talking survival."

They didn't like it one bit, but they couldn't budge Eric. Oh, he was unpopular. I was worried he was going to destroy what little pilot morale we had left or tie the company up in a messy strike, which would really finish us, but I had agreed to keep my hands off. For the time being I decided to leave them to fight it out on their own.

It was August and hot and I decided to get out of town for a rest. Glenys and I had property at the Indian Point end of Savary Island and had built a small cabin on it. The plan was that she would take the three kids up there—besides the two boys we now had a daughter, Marie—and spend a good part of the summer there while I would fly up on the weekends to join them. The catch was of course that I found it very difficult to get away from the office in those busy summer months, even on the weekends. By 1953 Glenys had taken enough of this and told me I had to do better—or else. I took her seriously for once and vowed that nothing would stop me from making it up on the coming Friday, about the middle of August. Then I promptly came down with a case of stomach cramps and nausea so severe I could hardly move. I didn't feel like eating, sleeping or living. I certainly didn't feel like going to Savary. But the thought of phoning Glenys to cancel out was even harder to face. I had made that phone call too many times already, for reasons not as good. So I packed my shaving kit and staggered out to the airport. During that day at work I began to feel much better, got through most of the work on my desk and made the flight up to Savary in good spirits.

By the time I got up the road to our camp the pains had returned. Glenys took one look at me and got on the portable radio phone to have a plane diverted to Savary at the first possible moment. Then she struck camp, did all the packing, bundling, roping—got the kids ready, and had everything on the beach within the hour, with no help from me. I was by now barely conscious. They took me home and put me to bed and fetched our doctor, Al Russell. He listened to Glenys' story, took a glance at me and called an ambulance. I had a ruptured appendix.

I spent the next five weeks mostly on my hands and knees in a bed at the Vancouver General Hospital, with drain tubes hanging out of my middle and other tubes going into my arms. In this posture I greeted armies of visitors and held regular meetings with my staff to direct the delicate manoeuvrings of our ongoing air-war. I received many get-well cards and letters from near and far but the one that impressed me the most was a very expensive bouquet of flowers and a card from Russ Baker.

By pure coincidence the hospital ward I landed up in was occupied at the same time by my partner in the radio business, Jack Tindall, who was in for exploratory cancer surgery. Neither of us were taking our problems too seriously and we were great company while it lasted. We were both confined to wheelchairs and on one famous afternoon got involved exploring the basement passageways under the hospital complex, which are like the catacombs of ancient Rome. We spent a good part of the day totally lost and finally got into an elevator which delivered us unexpectedly into the midst of the women's ward, not a popular move with the occupants. Jack got out of hospital about the same time as I did, in good spirits and grateful for the time we'd been able to spend together.

I recovered my strength slowly and went back down to the office, but I was pretty wobbly. Grant McConachie noticed this at one of our Western Aviation Council Meetings and right out of the blue said, "Jim, we're starting twice-monthly flights Vancouver to Lima. Why don't you take a holiday? The plane is half empty."

I thanked him sincerely but explained I just couldn't spend any more time away from the family—I had been away so much.

"Take 'em with you," he said.

So we packed our bags and forgot all about QCA for two wonderful weeks. In Lima I got talking planes with an ailing old stunt pilot named Bill Fawcett, who was owner of Peru's principal airline, and he gave me an open pass to fly all through the interior of the country, including stops at Cuzco and Machu Picchu, the Lost City of the Incas. I guess it was probably the first time in my life I had taken a real holiday and it did us all a world of good. The only unhappy note came when we received a cable from Vancouver telling us that Jack Tindall had died.

When Jack took ill we had put the radio business up for sale and Marconi offered twenty-five thousand, but they pulled out at the last minute. Poor old Jack, he'd been counting on it to take care of his family and he took it pretty hard. He succumbed before another buyer could be found. Now I had that business to worry about on top of everything else. Our old engineer Herb Hope had long since departed, but his wonderful family had yet one more rabbit in the hat. This was a sister, Winnifred. She had been working in the assembly plant and was now one of the senior people, and damned if she didn't take hold of things. I more or less had to appoint her acting manager in Jack's place. She didn't leave me any choice. When it came time to appoint a permanent manager she was an even more obvious choice. She turned out to be the best one we ever did have.

Meanwhile, back at the airport, more bad financial news. Even with the $150,000 subsidy, we had ended 1952 with a backbreaking loss of $278,000 and our revenues for the first half of 1953 were far below projections. Kitimat traffic was edging upwards but we still weren't doing any volume there to speak of. The greater problem, however, was that our non-Alcan revenues were down. The logging industry continued in a severe recession and our east coast service to Alert Bay and Minstrel Island was down to two flights a week from the previous seven, and even at that had load factors of only twenty percent. The west coast route to Tofino was slightly better, supporting two flights a week instead of the normal six. With all our Class 3 and Class 4 work subcontracted to Bill Sylvester of BC Airlines, we were only flying four planes regularly—two DC-3s and two Cansos, and it wasn't enough. I was quite sore at the Air Transport Board about this because it had been them who pushed us into it. The thought had been that we would get CPAL's mainline run to Port Hardy and maybe Terrace as well, but this hadn't transpired. We were permitted to fly the DC-3s into Port Hardy, but only to transfer through passengers to Alert Bay, Sullivan Bay and Minstrel Island onto seaplanes. We couldn't touch any local North Island traffic, which was still being rather awkwardly served by CPAL DC-4s and Convairs touching down on their way through to Sandspit. Worse, the Air Transport Board wasn't following through on its promise to reserve the "spine" business on the rest of the coast for us.

In the latter part of 1952, Baker had turned up with yet another Alcan contract, this one with Kitimat Constructors, to do all their flying between Kitimat and Prince Rupert and Kitimat and Vancouver. The Air Transport Board at long last was moved to take action. Order 1019 ruled that CBAC had indeed been operating illegal flights from Kitimat to Terrace and Kitimat to Vancouver all this time, just as we had repeatedly said. Baker, according to this order, was "prohibited from operating between Vancouver and Kitimat in either direction either directly or via intermediate points." Further, the new contract to fly everywhere for Kitimat Constructors was struck down.

I immediately wrote John Baldwin an appreciative letter saying, "I would like you to know...that every one of us here realizes and appreciates what you have done personally to bring this improvement about. Of one thing I can assure you—we have the personnel, the determination and, if permitted to do so, will provide all the necessary effort to make this an airline that you can be proud of."

The only trouble was, the new ruling had no noticeable effect. Baker kept flying full planes down the coast and we kept flying empty ones. Our Alcan traffic actually declined from nearly two hundred passengers a month to less than a hundred by the end of 1952 while CBCA's remained steady at a thousand-plus. Baker and Alcan nevertheless raised a great halloo against the board for daring to interfere with free enterprise. Baker screamed that "the Board has been and is sponsoring QCA." Alcan came to his defence, claiming that Air Transport Board interference was delaying the great project and they were having "inefficiency thrust upon [them]." This was a reference to our subsidy, which Baker had promptly turned into a major controversy with the aid of the *Sun* newspaper. He claimed that our problems were all self-inflicted and stemmed from the fact QCA was run by an "amateur." He would, he said, be happy to take over all our routes and fly them without any form of government aid. "Basically the problem resolves itself around QCA holding all the licences and the customers resisting their product while CBCA has more customers than can be handled and no licences."

But up at Alcan, it wasn't the customers who were resisting our product, it was their bosses. This led to some bizarre situations where people wanted to use us but the company wouldn't let them. On one occasion our Canso arrived on time at Kitimat and anchored out in the bay. Passengers began gathering on the beach. After all the talk we'd heard about the Air Transport Board and the cabinet ordering Alcan to install a proper base, we still couldn't get ashore on our own or take passengers in until the company brought a boat out to us. We weren't permitted to have our own boat. We still were not permitted to have an agent on the beach to give our pilots safe weather reports, relay schedule information to customers, take reservations or ticket passengers. We just had to wait till it pleased Alcan to come and get our passengers and bring the others out. In this particular case there got to be a real crowd of men all desperate to get out of camp that day, and the plane was due to leave, I think it was at two P.M. They were very restless, but Alcan just ignored them. Finally the men got so damned angry they started shouting and throwing things, and they raised such a ruckus that the office telephoned Prince Rupert and had the RCMP send two officers down in a Norseman to keep order. That happened more than once, when the police had to intervene to keep order among passengers who wanted to fly QCA and couldn't get aboard. Things were really Wild West and hot.

Toward the end of 1952 Baker applied for a Class 2 licence between Kitimat and Vancouver. He was constantly applying for routes he had no chance of being given, so we filed a routine notice of intervention and paid no particular mind. A Class 2 route of this scope could only be flown by large two-engine aircraft, which Baker didn't have. Then we heard he had bought a Canso fully outfitted for IFR night flying in anticipation of being given the licence. This was alarming. Under his present operating certificate Baker was firmly restricted to Group B aircraft, single-engine or small twin-engine craft with disposable load under six thousand pounds. It was a key part of our understanding with the Air Transport Board that he and all the other charter outfits would stay in Group B as we relinquished our small craft and moved exclusively into Group A. Baker's acquisition of an A-Group Canso along with his application for the Class 2 licence was a challenge to the whole plan to which we had now firmly committed the fate of the company, and I rather anxiously broached the matter with John Baldwin when I was in Ottawa on December 19, 1952.

Baldwin assured me Baker was buying the Canso purely on spec, and had received no assurance from the Air Transport Board he would be able to fly it. But to my great dismay I heard Baldwin now suggesting Baker's application for Class 2 to Vancouver along with Group A status might be granted. The aluminum company was putting formidable pressure on the government, and H.R. MacMillan had gone so far as to write a strong letter favouring Baker's application. Noting my expression of alarm, Baldwin sighed that Baker was now so firmly established on the coast that there might be no realistic way of stopping him. It was the first time I had seen him look defeated, and I realized the kind of pounding he must have been taking over this whole issue. With a sinking feeling I realized the game was lost. He had tried to stand up for the integrity of existing government policy, but we had evidently run up against a superior force.

Baldwin started musing that perhaps the whole Alcan mess had resulted from mishandling by the Air Transport Board in the first place. Perhaps the board should have refused to approve the low-rate CBCA contract and should have insisted that it meet authorized Air Transport Board rates. Did I feel the board had mishandled it, or else why the mess?

Of course, I thought all of that and then some. But he looked so damned unhappy I was afraid to agree with him lest he run out and jump in the Ottawa River. Absurdly, I found myself reassuring him that it probably didn't make much difference—Baker had got around the customer and probably would have got in there regardless of anything the board could have done. Possibly we, QCA, might have stopped it ourselves if we'd opposed his application for a Kemano base, but once again, I doubted it, I said. I immediately hated myself for uttering such blasphemy, but it seemed to cheer him up and after a while he started in again trying to convince me that giving Baker his Class 2 licence might not be such a bad thing if the aluminum company agreed to an equal division of traffic between the two airlines, which the board was going to insist on. When I left him I felt like jumping in the river myself.

The new scheme, when the board announced it on March 6, 1953, was a masterpiece of bureaucratic doublethink. Baker was awarded a Class 2 licence for service between Kitimat and Vancouver and so were we. Thus licensing, which had been devised to prevent competition, became a means of ensuring it. Baldwin avoided commenting on this reversal of the Air Transport Board's reason for being, simply accepting the aluminum company's claim that they needed more aircraft and saying "QCA is faced with certain other problems in its operation and should not be called upon to extend its service on this route." As if the worst thing you could do for a starving airline was fill up its planes! Under the new system both airlines were allowed three Canso trips per week into KKK territory and it was agreed the Alcan companies would divide their business equally between us. Baker was greatly alarmed at this prospect and at first rejected the deal, but Alcan's traffic manager, Lockwood, wrote reassuring him, "Even though they [QCA] are given the same rights as you, it does not necessarily follow that we will give them additional business." They had apparently decided to ignore the traffic-sharing deal even before it started.

For the first six months of 1953 we did in fact receive some additional business, with KKK passenger bookings climbing as high as 650 in June, but then Baker paid me the honour of one of his drunken phone calls to boast that he'd fixed things up, and the business abruptly vanished again. When we wrote the companies on June 12 asking what had happened, they just gave me more smoke to shovel. Kitimat Constructors said they could only make use of one flight a week from us, adding they might "decline surface transportation or accommodations" should we attempt any more than that. Morrison-Knudsen reverted to saying they were obligated to CBCA by contract, and "after having availed ourselves of the limit of Central BC Airways' facilities we have depended on them to contact other agencies to meet our requirements."

When I showed the board copies of their letters they said they "just couldn't understand it." I couldn't understand why they had ever thought it would be different.

At the same time I began to hear about Baker buying the Canso I began to hear about Baker buying out Associated Air Taxi, the charter outfit started by Bob Gayer. Gayer had done a surprising job with AAT, establishing a couple of small Class 2 licences to Pender Harbour and the Gulf Islands and pushing his revenues up to several hundred thousand dollars in 1952, but the operation finally collapsed under the weight of its debts late in the year. Bob and I talked about working something out between AAT and QCA, but we were staring bankruptcy in the face ourselves, and then when we got the subsidy we had to agree not to spend it on new ventures. Gayer needed a hundred thousand cash to pay his debts and he knew Baker was desperate for a Vancouver base that would establish him once and for all as a legitimate coastal operator, so he made an approach through Karl Springer. They agreed on a deal, but then Baker held off on finalizing it until March of 1953, thinking he could starve Gayer into bankruptcy and get the coveted Vancouver base for nothing. Somehow Bob held out, and ended up getting his hundred thousand plus a good swack of CBCA shares.

Baker now had his hand on two of the crucial ingredients he needed to complete his invasion of coastal flying — a legitimate Vancouver base and a Class 2 licence to fly large Group A aircraft the length of the coast. What we had been able to pass off as an overblown bush operation was now starting to shape up as a real challenge to our status as the BC coast's reigning regional airline. Our total of 6,101,037 passenger miles flown in 1952 was still well ahead of CBCA's 4,494,998, in spite of all our difficulties in that year. But 1953 was coming much closer, 6,421,818 to 6,363,174. It was in May of 1953 that Baker changed the name of the airline from Central BC Airways to Pacific Western Airlines.

By April we had gone through the last of the twenty-five-thousand-dollar subsidy payments from Ottawa and things didn't look good. Logging activity was still at rock bottom. KKK was an empty promise, our small planes were gone and our scheduled routes weren't producing enough to see us through the winter. We were receiving a lot of threatening letters from lawyers and rude phone calls from collection agencies.

Eric Bendall, meanwhile, had our operations department turned inside out. He had dusted off an old Link trainer and paid somebody about a month's wages to put it in shape. He was insisting on standardization of cockpits in all the aircraft, so that the cockpits of the two DC-3s and the two Cansos would become identical. As it was, all the planes had come from different backgrounds and had a lot of differences in the placement of instruments, radios and controls. Our pilots had never even mentioned it as a problem, but to Eric's way of thinking, it was unacceptable. Under stress a pilot might get mixed up and look at the wrong gauge. A valid enough concern no doubt, but the cost of standardizing the whole fleet ran into thousands of dollars. I argued for a recognition of priorities, given our desperate financial situation, but Bendall was inflexible. It was safety first and that was that. Every time I turned around he had people tied up in the Link trainer, out on training flights or sitting in classes. And he did less revenue flying than other operations managers we'd had. Staff productivity was down.

Equipment utilization was down too, because he wouldn't allow the slightest deviation from book procedure. When a plane came due for a maintenance check, it had to deadhead directly to base and a replacement had to deadhead out, even if this meant two non-revenue flights the length of the coast. Before, we would have put the check off till the next time the machine came through base on a revenue flight, but now any pilot who tried to stretch a point to save us some money could expect to be grounded, if not fired. Some of our most senior men got the heave and boy, there was some squawking about that. Ken McQuaig was about fifth in overall seniority and thought he could do no wrong. But when he landed a Norseman too hard in a patch of chop and collapsed a strut, Eric canned him. McQuaig sued us for wrongful dismissal and won. Eric didn't bat an eye. Another senior pilot was coming in late with a stretcher case and arrived in Vancouver half an hour after official grounding time. It was still light enough to see, but we had a policy of not flying after official grounding time for any reason. There was the usual clutch of reporters waiting with the ambulance and the next day the pilot's picture is all over the papers as a hero. Day after that, Eric cans him.

"May I just ask you what the reason was?" I said.

"Yeah, he should have delivered the patient to the hospital in Powell River before grounding time and remained there overnight. He flew on to Vancouver so he could get his name in the papers."

This was not a man who was trying to win popularity contests with his old flying colleagues. Or with his boss. Every time I turned around he had a new fistful of bills he wanted me to pay. We'd had "think airline" campaigns before, but this guy was something else. He was going through the organization like a dose of salts. It was what I'd always wanted, but coming at the time it did, I was afraid his cure would turn out to be the kind that kills the patient.

Baker was once again rubbing his hands in anticipation of our demise, and we began to read in the papers that he was buying us out.

It was time for another trip to Ottawa. I knew Baker was waiting for this so he could crank up his propaganda mill and try to queer any prospect we might have of getting government money, so I left very quietly via Seattle and took the US route, cutting north through Buffalo and Toronto. The first thing I did was go into the Air Transport Board, and I found everybody waiting for me. Dan MacLean and Romeo Vachon both laughed and said word had already got around that I was on my way back to lay the groundwork for another government handout. Baker's information service was better than I thought.

The mandatory visit to Tedford at the Post Office was even less encouraging than usual. Nothing doing on our most recent air freight proposal. Nothing doing on our proposal to add Comox to the airmail service. This, he said, was not the time to talk about increasing service, as the post office was in the red and cutting back. I took him out for lunch anyway.

Then came the only good piece of news I got from the trip: Carter Guest, the ageless terror of west coast aviation, had finally, after ten years of continuous appeal from every operator in the region, been transferred out of Vancouver. His replacement as district commis-

sioner of air services was to be a rising young star in the DoT named
H. Donald Cameron, whom the board assured me would waste no
time beating Alcan into line and bringing about a fair division of
KKK traffic. I tried not to laugh.

Ottawa was in a thoughtful mood, drifting and rather unfocused
in the shadow of the looming August 1953 election and with all the
political heavyweights out in the ridings mending fences. Everybody
I went to see felt like talking.

Both Jim Sinclair and Ted Applewhaite were out on the hustings,
but Jack Gibson had decided not to run in the election and was in his
office cleaning out files. He and his brothers had just sold their
extensive timber holdings on the west coast of Vancouver Island to
the East Asiatic Company and had been paid right off, so he didn't
need to be in Ottawa anymore. Jack was very fed up with the
Socreds in BC and with Canadian politics in general. "We were just
lucky," he said of his family's success on Vancouver Island. "No one
else was awake to the possibilities of the west coast at that time or
they'd never have let us get away with it." He said if they'd continued
and not sold out, they'd have been in serious trouble. The big money
boys, the H.R. MacMillans, would have closed in on them and
they'd have had it. Speaking of which he said he was very surprised
to learn that H.R. would prostitute himself by writing testimonial
letters for Russ Baker's licence application.

He felt we were facing an almost impossible situation. He said
Alcan was obviously too big and too powerful for even the
government to handle. Campney was against us, but he didn't think
Campney was sinister so much as simply misinformed by Baker. He
told me I could go talk to him if I wanted, but warned me not to
confide in him. In any case the key man for us remained Baldwin,
and he was on our side as much as he could afford to be. C.D. Howe
was still king of the Hill of course, and he was best got to through
Sinclair. I showed him a letter I'd written to Sinclair on the subject
of our subsidy and he agreed it was the kind of documentary support
that was needed, as Sinclair had been badly battered over the subsidy
by Vancouver Conservative Howard Green and Tory aviation critic
John B. Hamilton. This had all been triggered by Baker's media
campaign and according to Jack it had taken its toll on the whole
Liberal caucus. To have any hope of restoring the subsidy for a
second year I would have to wait until after the election and hope the
Liberals were returned, but without Campney.

He was in a reflective mood and kept skipping from one thought to another. "You know, there's something we should look at here," he continued. "H.R. MacMillan is a director and shareholder of the Union Steamship Company and he has more than a casual interest in the possibility of PWA taking over from QCA on the coast. This would give MacMillan some influence over coastal air services and their effect on the Steamship Company. And Campney acts for the Union Steamship Company as well as acting for Alcan and Baker. All of which adds up to a pretty solid interlocking of interests." In view of all this he wondered if we shouldn't just sell out to somebody — at the right price. He said maybe that time had come. Or the only other thing would be for us to retreat to high ground and operate a tight little air service at a lower cost and let PWA do what they will.

I had a few hours of unscheduled time, so I went over to the Department of National Defence on a prospecting trip. The prospect didn't pan out but it still proved to be one of the most interesting visits I'd made in Ottawa all year. The man I ended up talking to I'd known from the early days of the war when we'd taken him on one of our very first charters in the old Waco. His name was Doug Belyea and he was an old west coaster with connections to the well-known Vancouver merchant firm, Gordon Belyea. He was very bright, very well-connected and had risen to the top of the Ottawa heap. He had been keeping tabs on us, and he undertook to sit me on his knee and teach me the ABCs of the political situation in which we now found ourselves enmeshed.

He suggested we find out who our bank's major clients were and also talk to the bank directors. He said we should get the whole family tree of the bank's directors and figure out who we should get next to. He looked at our situation and decided we were presently in a very tight squeeze unless we could gain some powerful friends. Baker had them; we didn't. If we wanted to survive intact we would have to match him, otherwise our only option would be to find a buyer who wanted a ready-made chunk of the country's air transportation industry enough to put up some serious money, which might not be easy. He offered to set me up with introductions to the industrialist Frank Ross, the BC seed and feed magnate Ernie Buckerfield and a whole list of BC business leaders.

All this deep thinking got me going and on the way home I found myself trying to summarize where I thought we stood. The highlights of my report to the directors were as follows:

The good news first —

1. We have solved our immediate equipment problems with DC3s and Cansos.

2. We have no immediate need for capital expansion.

3. We have necessary operational growth and efficiency.

4. We have eight years' backlog of technical experience.

5. We have established top level liaison with government departments in Ottawa.

6. In Vancouver we have acquired all necessary ground service and passenger facilities.

7. We have established good public relations.

On the other side of the ledger —

1. We are temporarily short of revenue.

2. We need more substantial and diversified operational routes.

3. We need stronger political and big business support. Being small and independent we are defenceless against political and financial groups. We need friends at court. We need influential people on our board of directors. From here on, it won't be what we know but who we know that will make the difference.

FIFTEEN

Free Enterprise

I MADE SOME HALF-HEARTED STABS at getting to know some big-time cigar-chompers, and we flew Skeena MP Ted Applewhaite around his riding for free, but that's as close as I got to infiltrating the Canadian establishment. I just wasn't good at it and I more or less fobbed the job off on Jimmy Wells, the Ottawa lobbyist, whom we now retained as a kind of all-purpose advocate in official circles. Problem Number Two, getting more routes, was more the kind of thing I could sink my teeth into.

At this time we had the Terrace application coming up. Canadian Pacific Airlines had applied for it, Vancouver–Port Hardy–Terrace Class 1. Terrace had a full-length airstrip with a limited sort of RCAF radio navigation facility and would serve both Prince Rupert and Kitimat through a system of Class 2, 3 and 4 feeder operations. We intervened in the CPAL application with a proposal that we be given a Class 1 licence to serve Terrace and Kitimat using Cansos from Vancouver.

Now, what was PWA going to do about it? That was the question. They followed our example by intervening with a Class 1 application to serve Terrace with ten-passenger Grumman Mallards, which were not a Class 1 aircraft. It was obvious to everyone that they stood no chance. It was just a nuisance application to give them a bargaining position, but how were they going to bargain?

We talked to PWA and said look, if we get this licence to Terrace, we are prepared to let you do all the feeder service into Terrace. From Kitimat, from Kemano, from Prince Rupert — from the whole area. We'll pull all our small aircraft out of Seal Cove—where we still had a few floatplanes serving Stewart and the Queen Charlottes—we'll give you all the specialized work—just feed it into Terrace from our mainline run Terrace-Port Hardy-Vancouver. This was our proposal to them.

Just before the hearing there was a big ceremony out at the Vancouver Airport. They were opening a new runway or something and the airport manager, Bill Inglis, had all the bigshots out there—the mayor, the MPs, the cabinet ministers—I think even C.D. Howe was there. The centre of the activity was over on the north side of the field near the Air Force base. I had Hal Suddes with me and we were talking to Bill Inglis when Ralph Campney came up to me. He hadn't been defeated after all and he was still in the cabinet. His firm was still acting for PWA and he was towing Russ Baker behind him.

"I want you two to talk," Campney said. "Let's have an end to all this fighting that's been going on the coast. Now is your chance."

In front of Bill Inglis and all the rest that were there, Campney said, "I can talk to Howe and I will absolutely guarantee you, Spilsbury, if you and Baker will get together and oppose Canadian Pacific Airlines, they will not get the Terrace run.

"Jim, you've got the aircraft to fly the long route and Russ, you've got all these small aircraft—it's all you can possibly do feeding the thing—if you two can get together I'll see that CPA doesn't get it and QCA does. Now how about it?"

He turned to me. "What about you, Jim?"

I said fine, I'd been for it all along.

He turned to Baker and said, "Russ, how about it?"

Baker said it was fine by him, too.

Campney pursued it. "Russ, do you mean you will not apply on PWA's behalf but you will support QCA in opposing CPA, if Jim guarantees to give you the feeder work?"

"Absolutely," Baker said.

Campney made quite a ceremony of this, with half the aviation industry looking on.

"Let me see the two of you shake hands on it," he said.

And we did.

Baker said boy, would he ever be glad to see the end of all this infighting.

I left the ceremony elated. With our spine route through to the north coast, QCA was back on solid ground. I couldn't wait to get back to the office and celebrate. But when I got there I found I couldn't get anybody else enthused.

"You made a deal with *Baker?*" they said. "Jim, you'll never learn, will you? You can't trust that son-of-a-bitch."

Sure enough, when the hearing came up a few days later, Baker doublecrossed. He withdrew his opposition to CPA, did not get up and support us, and it turned out he'd made a deal on the side that McConachie would try and turn the feeder service his way if CPA got the route. Which they did. When we reapplied, PWA intervened against us again. Campney's law partner, Walter Owen, made a prolonged attack on QCA, describing us as "a company that depends for its existence on subsidies from the people of Canada, then wants to extend its service and put another solvent company off the run." He also took up the free enterprise cry. We were represented by Jimmy Wells. Some of his remarks are interesting in that they represent the response of an eastern air-wars veteran who was viewing the west coast situation for the first time:

> As most of you are aware, I am a visitor to the west coast. I am suitably impressed by your mountains, but it is your newspapers that have me agog! Down east an airline is happy if it gets a line or two once a month. But in Vancouver, I have been shown headlines two inches deep proclaiming "Air War Between Two Companies," and similar publicity is very common. What has shaken me is the failure of these air-minded newspapers to publicize the completely different manner in which the west coast has been treated.... I do not know of any region where communities are more dependent on air service; and yet QCA, the operator attempting to provide the essential service, has been compelled to shift for itself, and to face a degree of competition in excess of that faced by any other regional airline. In the east...if the community needed an air service but sufficient passenger traffic was not available...-mail pay took up the slack. Here on the west coast, QCA receives somewhere between one-quarter and one-third of what would be paid for a similar mail service in the east, and yet my learned friend talks of subsidies.

> The newspapers have a beautiful story, but they
> don't print it!

Wells was very impressive, but the Archangel Gabriel couldn't
have saved us against the combined opposition of CPAL and PWA.
We lost the application, and it was a real setback. In spite of all the
Alcan opposition and C.D. Howe and all the rest of it, if we had
been able to extend our Class 1 spine service into Terrace and
Kitimat, we would have had the airline back on a solid footing. As it
was, the Air Transport Board gave Terrace to CPA and allowed
PWA's parallel Kitimat licence to stand.

We had to find more business somewhere.

When Bob Gayer sold Associated Air Taxi to Baker he'd signed the
usual caveat promising not to start up again in the same business for
five years, so he found himself adrift. My acquaintance with Bob
dated back to the time he'd painted our Waco with soot, and I knew
he'd been involved in all kinds of big plans about setting up a chain
of small operators across Canada, contracting back east and all the
rest of it. He was very well informed, a very high-powered
individual, so I hired him. I thought he just might be able to give us
the kind of help we needed in Ottawa.

It turned out that under the deal he'd made with PWA, it was only
Bob who was restricted from going into business, not his wife
Louise, who was the general manager of the outfit, or his chief pilot
Slim Knight or the rest of his staff. They had been looking around
for a way to get going again in the charter business without bringing
Bob's name into it, because the charter side of the business had been
quite successful. The little short Class 2 routes to Pender Harbour
and the Gulf Islands and the Class 3 route to Port Alberni were
money losers and none of us were too unhappy to see Baker get
saddled with them. But AAT had made a good thing of charter, and
the thing about this charter business was, it existed mainly in the
form of the personal contacts which they still had. All they'd sold to
Baker was their licence and a few beat-up floatplanes.

I took a look at this and said well, fine, why don't Louise and Slim
Knight and all the pilots and maintenance people who are any good
come on over and join QCA? We still had our charter licence, we
had all the facilities, we had some small aircraft to start, and they
could carry on. We set it up as a separate company so we'd still be
technically in the clear with Ottawa.

One choice little item was their account with the BC Telephone Company. It was fourteen hundred dollars in arrears and Baker was telling them to go to hell, it wasn't his worry. We went to the telephone company and said we'll pay this providing we can continue using the number. They said that was fine. So now we acquired the old phone number of Associated Air Taxi; we acquired the pilots and the engineers and then we moved them into the old gate building at the entrance to our hangar. We painted a big sign as long as the building, it must have been eighty feet long: Associated Air Carriers.

The customers hardly knew anything had changed. When they wanted a little timber cruising, a fisheries patrol or a stretcher case brought in, they phoned up the same number and Louise's same voice would answer the phone. The result was that ninety-nine percent of the AAT charter business came our way and not Baker's. When he found out, he was apoplectic. Literally. We were told he was ordered to bed by his doctor, on the verge of one of his drinking heart attacks. And yet it was the way his lawyer had drawn up the purchase agreement that made it possible—it was just wide open. But Baker said this was the dirtiest, most low-down trick that anybody had ever played on anybody else. Three months later he was still screaming about it in the *Sun* newspaper, blaming the Air Transport Board because it had been them who suggested he buy Associated in the first place.

He couldn't take it like he could dish it out.

The board was very amused at the way Baker had fouled himself up, but nervous about our involvement. When I was in Ottawa they told me they'd had lots of trouble with AAT and thought at last they'd got that outfit out of their hair. They were very worried over any reappearance of Bob Gayer under any flag. Also they didn't wish to see us back in the small plane charter field. I was able to assure them that outside of owning the company, it was being operated independently of QCA, and they rather reluctantly accepted the idea. Bob Gayer stayed on and did us nothing but good, especially among the Ottawa mandarins who thought they didn't like him. Associated Air Carriers ran very smoothly for us and added a badly needed injection of fresh cash to our second-half revenues, as well as bringing some good new talent into the company.

Baker's revenge wasn't long in coming. He incorporated a travel agency called Coastal Air Services which advertised single-seat fares from places like Alert Bay, Minstrel Island, Tahsis, Tofino—all our main traffic points. They'd charter a PWA Norseman, send it up north, sell fares on a per-seat basis and send it back again. The fares they offered were well below ours — at this point Baker didn't care whether he lost money or not, as long as it hurt us, and boy, this did. According to our best estimates we lost over seventy-five-thousand dollars in passenger revenue during the 1953 season alone.

When I asked the board to put a stop to Baker's travel agency caper, they said they were very sorry, but travel agencies were outside their jurisdiction. I went back to Vancouver very frustrated about this and watched helplessly as our load factors fell lower and lower all over the coast. Finally someone had an idea. I don't know which of us it was, but it smacks of Bob Gayer. He really added some colour to our method of operation. We said alright, two can play this game. If the Air Transport Board really can't touch travel agencies, let's set up one ourselves and start raiding CPAL. Frank Griffith happened to have a dead company on the shelf called Wychwood Ltd. so we bought it for a dollar and set up in business as the Wychwood Travel Agency, advertising fares to Port Hardy at about sixty percent of CPA's rate. I think we flew one well-patronized trip, then the phone began ringing off the hook. The first call was from Barney Phillips, one of CPA's big wheels in Montreal. "Spilsbury, what the hell do you think you're doing? You can't get away with this!"

"It's being done every day out here, Barney," I said. "You're not keeping in touch." I explained to him that we were losing all our business to Coastal Air Services so we had to find something to replace it and we'd decided to take his Port Hardy business.

"To hell you will!" he shouted, and hung up, BANG!

The same day we had a sizzling telegram from the Air Transport Board commanding us in no uncertain terms to cease and desist all travel agency activity on pain of losing all our operating certificates immediately, and Baker got one at the same time. It took less than twenty-four hours once we twisted the lion's tail.

Baker couldn't tolerate the slightest reversal. We had been maintaining reasonably good relations with Kitimat Constructors through the latter part of 1953 and probably getting no more than our share of their business, while we still had only a small fraction of the total Alcan traffic. Baker lodged a protest with the board claiming that we were benefiting from *favouritism* in our dealings with Kitimat Constructors. When the board curbed his travel agency ploy he accused it of "Gestapo tactics."

Baker's propaganda campaign was reaching a furious pitch and must have been costing PWA a lot of money, but he was more prepared to spend money on that than improving his service and facilities. It was quite apparent to us that Baker was going all out to scuttle QCA by any means possible.

One day around this time our engineers were warming up a Norseman in readiness to take out a load of passengers and the engine began sputtering. They shut it down, traced the problem to fuel and found someone had put sugar in the gas tank. We then checked several other planes and found they'd been tampered with as well. The fleet was grounded while all fuel systems were cleaned out from stem to stern. We posted guards around our base after that, but it still wasn't very comforting to be flying out over Hecate Strait with a full load thinking of all the things on the plane our enemies could have sabotaged without us knowing.

As the 1953 season wore on it became very clear that we were not going to have the $150,000 we needed to carry us over the lean months of winter, and we began to become very agitated about our chances of getting another subsidy. We *had* improved our financial position, but in spite of our streamlined operation, we were still projecting a loss of $110,000. Keeping in mind the cabinet's threat to strip our licences if we didn't pull out of the hole, we tried desperately to raise the money through private channels. But difficult as that had been before, it was doubly so now that the Air Transport Board had granted parallel licensing over our main route. This had shaken the confidence of the banks as well as our directors, who were not anxious to throw good money after bad. Baker was very well aware that a second subsidy was our only way to survive the winter of 1953-54 and stepped up his efforts to focus debate on the issue of free enterprise versus government intervention.

I don't suppose Baker cared much more about the principles of free enterprise than he cared about the principles of fair play but his continuous chanting of the "freedom of enterprise" slogan identified him with the campaign CPAL was waging for the right to compete against Trans Canada Airlines and got him a lot of sympathy in the industry. I immediately ran up against this when I went back east in October for the annual AITA meeting and spent several days in Ottawa testing political waters on the subsidy question. Everywhere I went Baker had been there ahead of me, poisoning minds and closing doors. The Air Transport Board was fully aware of our growing predicament, but were very reluctant to endorse another subsidy because of all the political stink Baker had been making about the first one. In any event, if we were forced to go the subsidy route, they said it would likely be much less than the previous one. They brought up the equity question once more, saying it was necessary to see us put some invested capital into the company to renew everyone's confidence. I responded that it was the board who had destroyed everyone's confidence in QCA by granting Baker the parallel licence. Around and around it went. I left the Air Transport Board feeling I could no longer count on their strong support, and went to the Air Industries and Transport Association.

Here, things were even more dicey. Baker had tried to get the current AITA president, de Havilland Canada director Punch Dickins, to side with PWA by saying he would place an order with Dickins for five new de Havilland Beavers. AITA Executive Director Bob Redmayne told me Dickins was annoyed with Baker for trying to buy his support with five Beaver aircraft. He might at least have offered five Otter! Nevertheless, Dickins liked the free enterprise line. In the US, secretary of state John Foster Dulles was still doing his best to keep the Cold War alive and Senator Joe McCarthy was still inflaming hysteria about the survival of the free enterprise system, so Baker's sloganeering echoed a familiar theme. Free enterprise became such a contentious topic among AITA members they decided the only way to deal with it was to hold a debate at the annual general meeting in Toronto. They gave the event a lot of publicity and quite a few people were looking forward to getting in on the thing. Even the civil servants were saying, "Wait until the debate," as if the future direction of the aviation industry were hanging on the outcome.

I was very reluctant to get involved and it was only on the urging of Hal Suddes and others in the company who assured me that it was very necessary that it be done and there was nobody but me who could do it, that I found myself in the very embarrassing position of getting up and talking against free enterprise. Up to that point I had always considered myself a proponent of free enterprise. I would have been much happier if the debate had been headed "Government Control as Opposed to No Control" or "Control Versus Chaos." Anyway, it was set up that Baker would argue *pro* I would argue *con* and the members would vote on the result. I had to lead off and I was so shaky I practically had to be propped up by Suddes. I have no recollection of how long I spoke, possibly fifteen minutes or so. Nor can I remember my precise text, but during the Terrace application we had worked out a standard comeback to the Baker position:

In the aviation business we have lately been hearing a lot about private enterprise, freedom of competition, etc. I am hard put to understand why this line of reasoning is applied by some persons (and some newspapers) solely to air transport. All of us, of course, would like to have complete freedom; but most of us recognize this is not possible, if we are to co-exist with others.

Protection, control, franchises, licences have become so much a part of our economic life in certain matters that apparently some people fail any longer to recognize them as such, and when the same principles are applied to a relatively new industry, such as air transport, these people act as though the principle itself was revolutionary....Steamship routes, buses, taxis, public utilities of all kinds, mining developments, the lumbering industry—in almost every instance obtain exclusive licences, or franchises, or rights of some kind or other, and in return are required to do or provide certain things...

I could do this part by memory, and after a few minutes I began to find my voice and get into the meat of my argument. A licence to operate an airline, I said, is really a permit to go ahead and build a service around it. This needs a great deal of investment of time and money, facilities and equipment, public education, all leading up to the establishment of a sound operation. This was where the Air Transport Board control was very necessary. Competition, I said, results in price-cutting and over-flying which seems to be to the customer's advantage in the short term, but in the long term creates more cost, which ultimately gets back to the customer. Meanwhile, warring airlines are unstable, and this shakes the confidence of the bank and the investors, who don't know from one day to another whether their money is secure. Much-needed capital is driven away, retarding growth and undermining the standard of service. Competitive flying results in diminished safety and the general public finds itself the victim of unnecessary accidents. There shouldn't be competition just for competition's sake, I said. The matter should be decided by weighing the public interest, and the public interest invariably weighed in favour of control.

When I was done they called on Russ Baker to defend the free enterprise side. He wasn't present, and someone offered to see if they could go find him. This was in front of a big crowd. The meeting was very well attended. After about ten minutes the messenger came back and said that unfortunately Mr. Baker was indisposed. He was still in his room. Someone in the meeting called out, "Is he alright?" The messenger replied, "Oh, I think he is. It was a young lady that answered." There was a collective gasp, some strained laughter, and then they called on Karl Springer, but by this time Karl was so drunk he couldn't stand up and talk. After a good deal of muttering and shuffling, a PWA employee—I thought it was Dunc McLaren but he says not—got up and spoke. He called for free enterprise, which he described as regulated competition versus regulated monopoly. He said what he had to say, but didn't appear to have his heart in it. Then it went into committee and they had a vote on it. The result was supposed to be announced to the media, formed into a resolution and made subject of a lobbying campaign.

A day or two later back in Ottawa I met A.S. MacDonald, the executive director of the Air Transport Board, who was full of compliments about my talk. My moderation in presentation gained me many marks, according to him, while the entire PWA cause had fallen flat. In committee, members from right across Canada got up and spoke in favour of regulation and protection, and not one spoke in the negative. Even PWA spokesmen, when given the opportunity, did not introduce any opposition. Baker had stumbled into the meeting drunk and set upon MacDonald, accusing the Air Transport Board and MacDonald specifically of many nefarious acts against PWA. MacDonald blew up and told Baker his actions were entirely out of place; that he himself was not suitable for his position in the company or in the aviation business; and that his methods and his ethics were very poor. Baker apparently was taken aback by this open rebuke, which took place in front of almost everybody in Canadian aviation, and adopted a hang-dog attitude. He was so disgusted, he said, he was thinking of leaving flying. MacDonald said that would be an excellent idea, he was in the wrong place and there would be nothing but trouble ahead unless he changed his methods drastically. MacDonald confided to me he thought from Baker's look that he was on the verge of another heart attack, and maybe the sooner the better. I had never seen A.S MacDonald so outspoken and rather enjoyed it.

This was in October, 1953 and we went home and waited patiently for the Air Industries and Transport Association to come out with its pronouncement that Baker's so-called "free enterprise" option was officially rejected by the industry, and government's prevailing interventionist style was to be affirmed. This wouldn't be binding on anybody of course, but with the whole industry behind it, it would effectively counter the propaganda Baker had been putting out and set the stage for government action in our favour, either in the form of mail pay or a continued subsidy. By December we were once more on hold with all our creditors and having to plead with the bank in order to meet our payroll. Sinclair was again girding his loins to take our case to cabinet, but was waiting for the political climate to improve. I hopped the plane to Ottawa on the 13th and checked into room 604 at the Chateau Laurier. I was getting now so I knew my favourite rooms by number. The place was like home to me.

The first call I made was on Bob Redmayne of the AITA to see what had happened about the great debate. He explained that the committee had reached an almost unanimous conclusion in favour of my proposal but the whole thing had been suppressed. The directors had decided the association shouldn't publish or release anything about it. They didn't even want to answer the letter we'd written asking for the result. He said there was a lot of politics mixed up in the operation of the AITA, adding, "Baker stops at nothing. He tries threats and he tries bribery."

Redmayne went on to say that Baker made a great mistake by being drunk at the meeting and not talking to his cause. He'd isolated himself from the rest of the industry and left himself standing alone in his challenge to the Air Transport Board. Nevertheless he had some extremely powerful friends and had caused the Air Transport Board to lose face over their policy reversals on the west coast. In Redmayne's words, he had effectively "put the board on trial." He had also created a great deal of pressure for the removal of John Baldwin as Air Transport Board chairman. In October Baldwin had been promoted to deputy minister of transport so there was some question as to whether he would keep the additional Air Transport Board responsibility anyway, but Baker was making sure. The thought of losing such a powerful and strategically-placed contact was exceedingly worrisome.

I was very annoyed at the AITA for covering up my great debate victory but I found as I made the rounds that the results had got around and helped swing opinion in our favour anyway. A.S.

MacDonald told me the Air Transport Board was once again fully behind QCA. Dan MacLean, the head of civil air at the DoT was suddenly very anti-Baker and pro-QCA. Baldwin was completely fed up with Baker and Springer. Les Knight, a DoT investigator who had just returned from checking up on PWA, admitted to being a personal friend of Baker's but said PWA was on shaky ground, not only because of the very intemperate way they had gone after the Air Transport Board, but because of their prospects: their business in the interior of the province had gone flat, the Kemano job was nearly finished and the forestry contract was tapering off.

The timing of the debate could not have been better for our subsidy campaign. Whether my brilliant oratory or Baker's bad behaviour deserved credit, it provided us with the break we were waiting for. Jim Sinclair went to cabinet with a proposal for a $125,000 grant and came away with a privy council order for $20,000 a month for the next five months. QCA was assured of living one year more.

SIXTEEN

The Tight Little Airline

BACK AT VANCOUVER AIRPORT, a surprising transformation was taking place. Eric Bendall was working wonders in our operations department. After the first few months we all began to realize that Eric knew exactly what he was doing. At long last, here was the man with the knowledge and uncompromising standards to mold us into the tight little airline we always wanted to be. We had been trying to establish IFR standards for at least three years but when Eric came on board at the beginning of 1953 Bill May and Art Barran were still the only two among our key staff with valid instrument ratings. Within ten months Eric had all our Class 1 flight crews and planes certified for IFR—even the Cansos—and the DoT granted us full IFR certificates on our main routes.

This had two immediately beneficial effects. We could now fly in bad weather. Cancelled trips and the lost revenues they entailed became rarer. The improved regularity of our flights helped build traffic. We could also fly at night. The early grounding times which had cramped our flight planning for so long were suddenly eliminated, giving us a much longer flying day and allowing us to design more convenient schedules.

From this point on Eric received very good cooperation from nearly everybody. He showed us what airline thinking really was. He raised us all, management and operations both, to a level of professional flying excellence that probably wasn't surpassed in any other airline our size the world over. And in so doing, he completely restored our confidence in ourselves.

In 1953 Baker drowned six passengers in a Goose, lost another planeload in a Widgeon and several more in a Strannie. In 1957 he crashed a DC-3 at Port Hardy killing fourteen people, but after Eric came aboard QCA never lost another passenger. Our Class 1 service chalked up 1.25 million miles without an accident, which was well above the international average.

The disapproving letters we used to get from the DoT and the ATB turned to letters of praise.

PWA's Port Hardy crash provides a glimpse of what Eric brought to the operation. The elevators on a DC-3 have to be fastened down in some way so they don't flap in the wind during stopovers. Many operators just dropped a small wooden chock into the elevator to jam it tight. With our DC-3s Eric insisted on using a cumbersome clamp affair with a chain going down to a heavy weight on the tarmac. This was immensely more awkward to handle, but Eric claimed it was safer. A wedge, he argued, could be overlooked during the pre-flight check and mistakenly left in during takeoff, while these chains would pull themselves off when the plane moved. Some pilots would argue that such hyper-caution was unnecessary because when the elevators on a DC-3 were locked the control column was jammed so far back you could hardly squeeze into the cockpit to fly the plane, but that didn't convince Eric. He was there for that one-in-a-million combination that would add up to trouble. Jack Crosby, the PWA operations manager, opted for the handier wedge type of chock, and sure enough one of his pilots managed to take off with locked elevators and crashed, killing himself and thirteen others.

The DoT was so impressed with the job Eric did for us they hired him away and eventually promoted him to Superintendent, Air Operations and Inspection for all of Canada, where he worked monitoring everything that flew until he retired in 1977.

With Eric's revamped flight operations ticking over like clockwork and coastal logging activity enjoying a resurgence, 1954 began to shape up as a good year — at least in comparison to the kind of years we'd been having. We were able to project a small profit for the first time since Mount Benson. We still had a large bank debt hanging over our heads but we were able to take care of our most threatening creditors and I began to feel we had come through the worst of our trials. But it was another megaproject that really turned our fortunes around.

I first heard about the Distant Early Warning (DEW) Line at the

AITA meeting in October, 1954. It wasn't official. The official agreement between the US and Canadian governments wasn't signed until the following month and it was still top secret. Bob Gayer was with me at this meeting, and the two of us jumped on the rumour. We got referred to one general, then the next. But the more we heard, the more excited we got. This was big. The US government was going to pay the cost of building twenty-two primary radar stations stretching five thousand miles across the Arctic and everything was going to be flown in—by Canadians. That was written right into the deal, all work including the flying had to be done by Canadians. At $350 million the project was even bigger than Alcan and its flying requirements would be greater by far. Tommy Fox, who operated Associated Airways out of Edmonton, was given the major contract for moving freight into the western Arctic, CPAL took care of the central portion, and Maritime Central had a contract for the Atlantic side. Each of the major contractors subcontracted out portions of their territory and we were invited into the Associated sector. Fox was buying planes like they were going out of style. He'd had just one large plane, a Bristol Freighter, but now he bought four Avro York ten-ton freighters, two DC-3s and a Lockheed 14.

With our little fleet consisting of two DC-3s and two PBYs we couldn't spare any equipment without forsaking our own routes, and yet there was obviously going to be far more money in the DEW Line contract than all the rest of our stuff put together. The amount of tonnage they were talking about was almost frightening. I knew we were in no position to buy equipment, especially the big stuff, so Gayer, Hal Suddes and I brainstormed around and came up with the idea of getting the Flying Tigers. This was a group originally formed inside the US Air Force to fly supplies over the Burma Hump during World War II. They put together their own company after the war and had since become one of the leading air freight contractors in the world. But they wouldn't be able to bid on the DEW Line directly because they weren't Canadian. We figured we could cut some kind of a deal with them and I caught a plane to Hollywood to meet the head Tiger, Fred Benninger.

He was all for doing something. We worked out a deal where they'd lease us planes all painted up in our colours and send us crews who'd be on our payroll, in return for a cut of the gross. He was a very easy guy to make a quick deal with. We didn't even have a formal agreement for the first six months. We had two C-46s and one DC-4 freighter at the DEW Line base in Hay River in a matter of weeks. I sent Bob Gayer up as on-site operations manager and he went to Whitehorse and put his mukluks on and did a damn good job.

We got the Tigers' best pilots; the only thing was, none of them had any Arctic experience. I can remember when the first contingent arrived. A DC-4 went up to Edmonton Airport and everything was frozen up tight. I went around with these guys to tell them what kind of clothing Canadians wore in the north and they had a wonderful time. The Hudson's Bay clerk gave them these little red caps with peaks and ear flaps. They didn't look like much but in fact they were as good as the big furry ones and in an aircraft much more practical. I remember this big southern Californian trying on this little red cap and drawling to his buddies with a cockeyed grin, "Waaal, it don't look too good mebbe but that there fella swears it's garr-an-teeeed for forty below!" When we got them up to their hotel rooms we found that in the dry air, or for whatever reason, there was a terrific jolt of static electricity when you first touched the doorknob. A regular little lightning bolt about two inches long would appear. I showed them how to ground it harmlessly by touching their room key to the doorknob, but they didn't like that, they wanted the lightning bolts. So I showed them how to run up and down on the carpet scuffing their feet and build up a good big charge. You could really hurt yourself if you worked at it. These characters couldn't get over it. Half the night you could hear them hooting and playing with this. They were like kids out of school. I really wondered what kind of a bunch of galoots I'd got myself hooked up with.

But when it got time to haul freight they were all business. They brought all their own maintenance equipment and spare gear — everything except engine heaters. You had to pre-heat the engines in the morning by putting a big tent over them and giving them hell with one of these Herman Nelson space heaters. It takes four for a DC-4, one under each engine. Anyway, we bought a bunch of Herman Nelsons for them and they were away. Boy, did they ever tuck in and run up the ton-miles. Our freighting team really shone in the early going, and more and more of the business came our way. After the first few months we must have had ten big C-46s and DC-4s going full tilt. Nobody on the whole DEW Line hauled more tonnage except CPAL.

This was bringing in more money than we'd ever dreamed of, quite literally. I just couldn't *believe* the revenue figures we were chalking up. In 1955 I remember looking at an interim balance sheet for one month's operation and seeing a net profit before tax of ninety-one thousand dollars on a projected gross for the year of eleven million. This was over five times what we'd earned in any previous year.

The coast routes still were just running at break-even but with the DEW Line profits pouring in we were getting our fuel bill in line and paying down our bank loan at a rapid rate. It was just a matter of time before we had the airline back in the black. With cash in the bank and our new operational prowess, things were suddenly starting to brighten up. We would soon be in position to add some new equipment and expand our routes. Sinclair lent his support to CPAL's proposed polar route to Europe in exchange for a McConachie promise to cede us Port Hardy.

For Baker on the other hand, it was starting to look like the party was over. Alcan construction was rapidly winding down and would cease altogether during 1955. At that point he would be back almost where he started. He would still have his half-of-a-Class 2 licence serving Kitimat–Vancouver and his old licences out of Prince George, plus the very limited Class 2 and 3 charter licences he'd got from his purchase of Associated Air Taxi, but it was chicken feed compared to what he'd had under Alcan. Most of the traffic coming out of Kitimat would go to CPAL, which had the licence to fly land planes into Terrace. Baker's revenues would fall back to nothing. It was doubtful he could survive with the overheads he'd taken on, and even in his best year, 1952, he hadn't made over $32,740 net profit. He hadn't paid off his $375,000 debt to the IDB and in 1954 the company lost $51,104. He was making a start on the DEW Line but he'd missed the boat up there. He hadn't found out about it until we were already at work and all he'd been able to send up were some Ansons and Beavers. Later he leased a C-46 but his crew couldn't get on track. They blew engines, missed appointments and generally got themselves a reputation as bunglers.

We, meanwhile, were humping big tonnage. Back on the coast we had managed to come through the crisis with our important licences intact and we were poised to ride into the fifties with our dominant position on the coast reaffirmed. Baker's publicity campaign against our subsidy had succeeded — Jimmy Sinclair told us confidentially that there was no way he was going to get it reinstated for another year — but we had offset that by finally obtaining a tentative agreement with the post office for a mail pay contract worth

$125,000. At last. It was hard to believe, but one day Turnbull just said alright, he'd give it to us. What they had finally agreed to do was pay us the equivalent of one full-fare passenger on every regularly scheduled flight, up to two hundred pounds. Over that there would be additional payment at freight rates. It would still leave us far below the national average, but it was an eight-hundred-percent improvement over what we had been getting. The important thing, though, was the DEW Line contract, and we could see this was going to continue for three or four years and bring in millions upon millions of dollars.

Baker was getting desperate. In March I got a letter from Springer bringing up the question of "consolidation." The thought appalled me. If it were up to me, I'd see Springer in hell before I'd let him and Baker get near QCA. The trouble was, it was no longer up to me. I had been selling handfuls of shares to Lando and Griffiths every time I got up against the wall and I was now a minority stockholder. And much as I hated to admit it, I knew in my heart that if Springer offered them a chance to get out of QCA with a decent profit, they would take it. My only hope was that Springer and Baker would never be able to bring themselves to offer QCA what it was worth.

Not long afterward I got a call from Baker. He hailed me like a long lost buddy, as if the past four years had never even happened. He wanted to come down to our home for a visit.

"I want to see your paintings," he said. We were living on Drummond Drive in the Point Grey area then and he'd heard that I was a spare-time painter. He dropped around in a great big Cadillac. He always drove the biggest car you could buy and lived in the highest house. He had Madge with him and his daughter, they came in and oh, they were nice as nice. Finally when we're down in the basement looking at my pastels he puts his hand on my shoulder and says, "Look. You're a talented guy. You're the smartest radio man I ever saw. Why don't you just get back to what you're good at? I'll give you all the money you'll ever need for QCA. You're not cut out to run an airline, but I am, I've flown all my life."

Then he looks sort of wistful and says, "You know, if I don't make it in the next two or three years I never will, because my heart'll go on me. I got a bum heart, you know." He had had his first heart attack in 1953.

"So whaddya say? You can paint away all you want, you've got your little radio company, and you'll have all the money you ever dreamed of."

I smiled.

Then it was back to the dogfight. I think our board made some kind of counter-proposal where basically Baker's outfit would merge into us and we'd give them back a bit of stock or something; probably it wasn't meant to be taken seriously but it kept the pot boiling. Springer replied they couldn't make any deal where they didn't end up in control and Baker didn't end up chief executive. Baker was going around telling everybody he was on the verge of taking us over. Privately he was moaning to his friends about how unreasonable we were and how bad things were looking for PWA. He wrote C.D. Howe, whom he was on quite intimate terms with, and told him what a hard time he was having trying to get QCA to see the light. Howe wrote him back September 30, 1954, sympathizing and saying he had taken "one or two steps behind the scenes that may be helpful to your situation." At the same time, according to Condit, Baker sent one of his undercover men in Ottawa twenty-five thousand dollars in unmarked bills. It's interesting to have it documented, but I have always assumed Baker was bribing people left and right all along with cash, kickbacks, free trips, women—anything he could get them to take. That was his method. Condit quotes him at one point saying, "Every man has his price. With some it takes money, with some it takes booze, with some it takes women, and occasionally it takes all three."

Every day people were phoning up and saying hey, what's this about you guys selling out to Baker? How can you do this? Benninger of the Flying Tigers phoned me up and said, "The minute that SOB takes over, our planes are gone. We won't touch him.' I assured him he had nothing to worry about, it was the farthest thing from my mind. Eric Bendall came to me and said, "Whatever you do, don't make a quick sale. Baker is obsessed with taking over QCA and he'll pay any price. Keep him waiting and you'll get far more than its worth." A cash offer came in from PWA but it wasn't in the picture. Lando and Griffiths sat down with Baker and his advisers for some head-on negotiations but they couldn't see each other.

To help us make up our minds, the bank called and said we would be getting a final offer from PWA within a short time and they were taking a more than passive interest in the outcome. And just in case we didn't get the hint, they called our $400,000 loan on twenty-four hours' notice. I called up our local manager, wanting to know what was going on—it was the first time in years our account had actually been in decent shape—and he didn't know a thing. It had all been done over his head, from back east. I went over to good old Grant McConachie and told him what had happened and he went through the roof, "Don't let them get away with it! You go back and tell them they can seize the goddamn planes, Jim, and I'll give you all the equipment you need to keep going until you can get clear of the bastards!" They had pulled the same thing on him once and he'd been trying to get back at them ever since. So we went back to them and said go ahead, the planes are yours. We don't need 'em. Our loan was reinstated. Lando and Griffiths gave me an announcement to make that merger talks had failed and QCA was not for sale.

According to Condit, Baker went to his banker at the Industrial Development Bank and said, "We've had it."

That was fine with me. As far as I was concerned we had been through our trial by fire and now we were poised to reap our reward if we only waited for the pieces to fall into place. I felt no matter what price we got Baker and Springer up to, it would only reflect our past and not our future. And as far as I could see, whoever came out owning QCA would have the future of flying in western Canada to do with as they wished. Excepting CPAL, the only outfit of comparable size west of Ontario was Tommy Fox's Associated Airways, and from what we'd seen in the Arctic we knew by this time that Fox's outfit was coming apart. He'd taken on too much at once and he couldn't keep his planes in the air. He was cracking them up one after another. The whole west would just naturally fall to us as CPAL moved on to fulfill its destiny as a national and overseas carrier, and within a very few years QCA would be worth far more than anyone could see by looking at its book value in 1955.

But Lando and Griffiths weren't aviation visionaries like my cousin Rupert. They were businessmen, and being businessmen, they knew that when you have a shaky investment that takes an unexpected jump in value, the sensible thing to do is unload it fast. I could see how the land lay and I felt there was nothing I could do about it. I couldn't buy them out myself and I didn't know anybody else I could turn to. Anyway, I was too tired to think about it. I kept myself busy running the airline and left the negotiations to them. Eventually they got PWA to make a decent offer, they upped the ante, PWA came back with more money, and in July of 1955 the airline was sold. The final price was 1.4 million, of which my share was just over $400,000.

Baker naturally did what he could to make the changeover as

painful as possible. Part of the deal was that I was to continue in a consultative capacity at five thousand dollars per year. This was largely to enable me to complete the $125,000 mail pay deal I was in the middle of finalizing with the post office, but Baker never once called upon me and I could do nothing on my own.

A few months after the sale about four or five of our ex-pilots came down to my office at the radio company to ask my help. They were desperate. They said PWA was totally disorganized. Bendall had been transferred out. Under Baker's flight operations manager, they said, the operation was reverting to bush standards. Like Russ, he was a drinker, and their operational meetings often turned into boozing sessions. Most of the ex-QCA supervisors had been demoted in favour of PWA types. They said safety standards were being allowed to decline. Could I, they wanted to know, maybe intervene through Ottawa in some way? I told them there was just nothing I could do.

Apart from that one delegation, very few of our old people came around to see me. My company car, a Studebaker land cruiser, was left to sit in my old parking stall outside the PWA offices until the tires went flat and the interior mildewed, a symbol Baker seemed to enjoy. I offered to buy it at the full factory price, but he wouldn't consider it. He put a card in the window saying this *was* Spilsbury's car. I tried to tell him about the mail pay money going unclaimed, but he wouldn't meet me. Finally I got hold of his assistant, Dick Laidman. He knew nothing about the post office deal. I told him it was all in the files under "Air Mail" and he could look it up. He then told me that three days after the takeover Baker had had all the QCA files taken out behind the hangar and burned, an infantile act that cost the company—and future aviation historians—an irreplaceable resource.

I was eventually forced to sue to collect my five thousand.

Some QCA loyalists, like Hal Suddes, quit as soon as the sale went through, and Baker began stalking those who had stayed in the company. Hepburn was canned at the first opportunity, reputedly because he dared to show up in the parking lot with a Cadillac as big as Baker's. Poor old Hep died a few years later. Bendall was shunted over to run the Alberta side and left to join the DoT a year later. Gayer he kept for awhile and then ceremoniously beheaded in a company meeting after accusing him of a long list of imagined offenses. Slim Knight was cut down in a similar manner. But many of the ex-QCA staff were too difficult to do without and some, like

With the coming of the DC-3's and Eric Bendall's disciplined IFR operation, we were admitted to Vancouver's main terminal.

Flying Tigers' DC-4 awaiting DEW Line duty in Edmonton.

Flying Tigers' C-46 on the job at Hay River.

The DEW Line work went pretty well, but as Charlie Banting might have said, "You can't have everything absolutely perfect."

Dick Lake, our last superintendent of maintenance, and pilot Jack Miles, went on to take up key positions in PWA's executive suite.

For years I assumed PWA bought QCA with Karl Springer's money. It wasn't until I talked to author John Condit during the researching of his book on PWA in 1976 I learned Baker had bought us, in a sense, with our own money. Some $400,000 of the purchase money came through the federal government's Industrial Development Bank, courtesy of the airline provisions my AITA committee had been largely responsible for putting in place but QCA was never able benefit from. He didn't get all he needed for the buyout from the IDB but he got enough to put the deal together. He made the most of his friendship with the western director of the IDB, but the bank knew Baker had support where it counted. Our old friend H.R. MacMillan kept Baker under his wing all the way. We don't know what all C.D. Howe attempted to do "behind the scenes," but his influence in business and government circles was virtually limitless at that time. Springer was no slouch either. The fact was, Baker's big-name backers came through for him in the end. They weren't about to see their boy go down and they found a way to make it all work out to their satisfaction. Their friendship and support were Baker's only assets in the end, but it was all he needed. Doug Belyea was right when he told me you can't beat them unless you get somebody just as powerful behind you. There was no winning against these chaps. Even when you came in first, they gave the prize to their own candidate.

This was my loss of innocence in business politics. I won't say it made me into a cynic the way Baker was a cynic, but it left me with a healthy skepticism about the way things work, especially in the upper storeys of the Canadian business world. There's all kinds of room for the small entrepreneur to get started, roll up his sleeves and build a profitable small business. But when you go beyond a certain point and start crowding the big boys, you soon find a different set of rules coming into play. I always said I was doing okay until I got my gross up over a million dollars a year: then I became a target. They were such a small group in this country then, if you got on the wrong side of them, they could cut you off at your bank, they could tie you up with red tape, they could get you coming and going.

Looking back from my present vantage point, I find I have less complaint with the move my partners forced upon me back in the summer of 1955. PWA had offered a price they couldn't refuse and I couldn't have asked them to turn it down on my account. I pocketed a good slice of it and promptly went out and bought myself a fifty-foot cabin cruiser, in which I have since spent some of the very happiest days of my life, cruising up and down the BC coast taking time to paint some of the scenes I never had time to look at twice when I was rushing up and down the line with my head full of airstrip meetings and route plans.

From 1955 onward I devoted myself full-time to the radio business, just as Baker suggested, God bless him. The only flaw in the plan was I didn't get to retire properly until I sold Spilsbury Communications in 1981, when I was seventy-six. And while I escaped the airline episode with my health and sanity more or less intact, my marriage didn't survive. I was far too preoccupied with flying for far too long, and Glenys had a damn rough time. I wasn't home nights till late, working in the office, away all the time travelling, coming home dog tired — she was very glad when I sold. She said, "Now maybe we can settle down." But it didn't work out. We just seemed to grow apart and in 1957 we separated. It was perhaps the airline's most traumatic legacy in terms of my personal life, but time heals all wounds, and in 1970 I was married a second time, to Winnifred Hope, our long-time manager at Spilsbury and Tindall.

Spilsbury & Tindall may not have prospered but, with many ups and downs, it grew. I invested $100,000 of my proceeds from QCA in it, spent another $100,000 on a new factory, brought staff up to an average of seventy-five employees and realized about the same annual gross sales we had in QCA with two hundred on staff. We designed and built some very sophisticated single side band (SSB) radio-telephone equipment and sold it in Europe, Africa, South America, Cuba, the Orient — all over the world. To accomplish this, Win and I did a lot more overseas travel than I ever had in the airline business, but it paid off. By the time we sold the company it was the largest exporter of radio-telephone equipment in Canada.

Nor did I see the end of my habitual visits to Ottawa. I found myself still deeply embroiled in struggle with the DoT — the radio division this time — and ended up helping found the Western Canada Telecommunications Council to carry on the east-west tug-of-war on a permanent basis. In 1982 the WCTC honoured me with a special award to be given annually in my name — not to the candidate who puts up the best fight with the DoT, as it perhaps should have done, but to the one who contributes the most to marine safety through the use of radio. But in 1988 something even less likely happened. The DoT's successor, the DoC, actually *paid me* to make a trip to Ottawa, and when I got there communications minister Flora MacDonald presented me the first ever Communications Canada Technology Award. It was one of the pleasanter surprises perpetrated on me in that city.

The tight little airline I started by accident back in 1943 did pretty well for itself, too. Armed with our licences, Eric Bendall's airtight Class 1 operations department and our lock on the DEW Line business, QCA did go on to conquer the west, but under PWA colours. It did more than that: thirty years later it opened its mouth wide again and this time swallowed CP Air, becoming the second largest airline in the country. But none of this did any good for Russ Baker, who succumbed to heart failure and died at age forty-eight after only two years in my old hotseat. Bad living hurried his demise, but the pressures of the job would have taken their toll on whoever did it. Grant McConachie's heart also gave out under the strain at age fifty-six. If QCA had continued under its own name with me at the head it is hard to imagine I would still be here, painting my pictures, continuing my voyages of discovery up the coast and writing books damning my old enemies at age eighty-three. I doubt that I would have had the opportunity to take the trip I did in the summer of 1988, flying over our old Stranraer route at an altitude of sixty-eight thousand feet and a speed of one mile a second in the Air France Concorde. I remember noticing from the brochure that the Concorde had almost the same wingspan as the Stranraer, except on the Strannie we had it arranged in two layers...

Even if I had stayed on at the controls of QCA and survived the strain for any length of time, I know I wouldn't have enjoyed the next chapter in the airline's history, which involved expansion to the north and east with diminishing emphasis on my old stomping ground along the coast. It was the coast that inspired me from the first: originally, to bring it into communication with the outside world; next, to bring it into actual contact with the outside world through quick and affordable air transportation. I would have accepted the challenge of the national and international market, as I did in the radio business, but I don't think my heart would have been in it.

I was deeply disappointed to have left the airline when I did, but I always felt we left QCA in good shape. Condit, in his history of Pacific Western Airlines, claimed "total victory" for Baker, but I don't see it that way at all. In our actual head-to-head struggle for survival, QCA won and PWA lost. By the beginning of 1955 PWA was a spent force. Its rag-tag collection of small aircraft was superfluous to normal flying needs on the coast and it had a huge debt with no apparent means of paying it off.

Having lost the battle on merit, Baker had to find big money backers and buy what he once thought to win, and he had to pay dear. This was not competitive victory, it was outright purchase. You can purchase anything your heart desires if Daddy is rich enough. Meanwhile I felt my plodding commitment to solid service had come out on top of his flim-flam-the-public-and-win-over-the-boss approach. Our belief in operational soundness had allowed us to weather the misfortune of Mount Benson and finally paid off when under Eric Bendall's leadership we were able to stabilize our coast business and position ourselves to maximize DEW Line opportunities, while Baker's claims of superior flying smarts were revealed as false.

Although I still cringe to think of all the unnecessary grief it caused so many people—someone counted seventy-five deaths attributable to competitive flying during the Alcan project—I was never able to feel anything but proud about the way our great confrontation turned out in the end.

The most unpleasant part of the sale to me was to be suddenly cut off from all the people I had been working with and who had put their heart and soul into making QCA a success. I felt I was deserting them at a time when I felt I should have been right in there hitting hard to make all our dreams come true. I know that many of our employees had the same reaction. They figured I had thrown in the sponge when I was actually in a better position to carry on than I'd ever been. Time papered over many of the divisions of the Baker years, but a PWA management study reported after the sale there were still situations where old QCA employees wouldn't share information or take orders from old PWA employees and vice-versa, and long-submerged resentments can make for edgy conversations among us hoary survivors to this day. In 1984 we staged a QCA reunion party on the fortieth anniversary of QCA's founding and over two hundred people turned up. It was traumatic, but wonderful.

Index

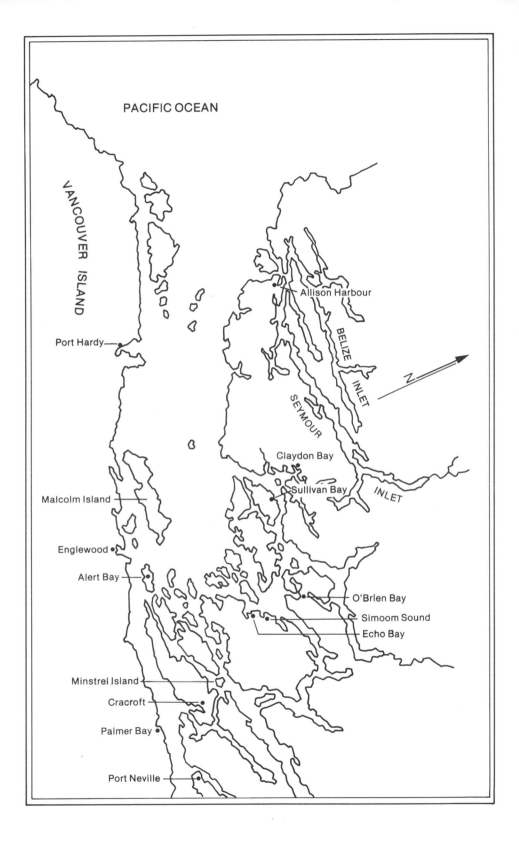

PACIFIC OCEAN

VANCOUVER ISLAND

Allison Harbour

BELIZE INLET

Port Hardy

SEYMOUR

Claydon Bay

Sullivan Bay

INLET

Malcolm Island

Englewood

O'Brien Bay

Alert Bay

Simoom Sound

Echo Bay

Minstrel Island

Cracroft

Palmer Bay

Port Neville

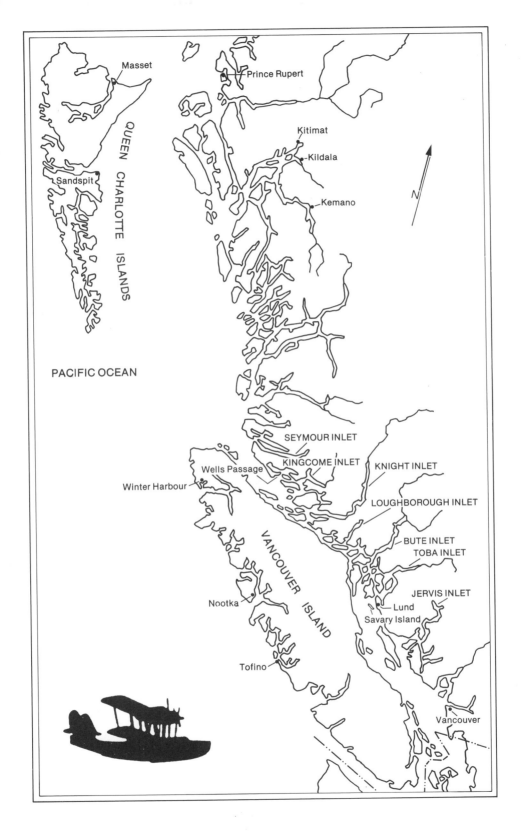

Masset

Prince Rupert

Kitimat

Kildala

QUEEN CHARLOTTE ISLANDS

Sandspit

Kemano

N

PACIFIC OCEAN

SEYMOUR INLET

KINGCOME INLET

KNIGHT INLET

Wells Passage

Winter Harbour

LOUGHBOROUGH INLET

VANCOUVER ISLAND

BUTE INLET
TOBA INLET

Nootka

JERVIS INLET

Lund
Savary Island

Tofino

Vancouver